AJ Sadler

Mathematics Applications

Student Book

Unit 3

T0359567

NELSON
A Cengage Company

Mathematics Applications Unit 3
1st revised Edition
A J Sadler

Publishing editor: Robert Yen
Project editor: Alan Stewart
Cover design: Chris Starr (MakeWork)
Text designers: Sarah Anderson, Nicole Melbourne,
Danielle Maccarone
Permissions researchers: Catherine Kerstjens, Helen Mammides
Answer checker: George Dimitriadis
Production controller: Erin Dowling
Typeset by: Nikki M Group Pty Ltd

Any URLs contained in this publication were checked for currency
during the production process. Note, however, that the publisher
cannot vouch for the ongoing currency of URLs.

For product information and technology assistance,
in Australia call **1300 790 853**;
in New Zealand call **0800 449 725**

For permission to use material from this text or product, please email
aust.permissions@cengage.com

National Library of Australia Cataloguing-in-Publication Data
Sadler, A. J., author.
Mathematics applications unit 3 : 1st revised edition / A J Sadler.

1st revised edition
9780170394994 (paperback)
For secondary school age.

Mathematics--Study and teaching (Secondary)--Australia.
Mathematics--Problems, exercises, etc.

Cengage Learning Australia
Level 7, 80 Dorcas Street
South Melbourne, Victoria Australia 3205

Cengage Learning New Zealand
Unit 4B Rosedale Office Park
331 Rosedale Road, Albany, North Shore 0632, NZ

For learning solutions, visit **cengage.com.au**

Printed in China by 1010 Printing International Limited.
9 10 11 12 26 25 24

PREFACE

This text targets Unit Three of the West Australian course *Mathematics Applications*, a course that is organised into four units altogether, the first two for year eleven and the last two for year twelve.

This West Australian course, *Mathematics Applications*, is based on the Australian Curriculum Senior Secondary course *General Mathematics*. At the time of writing there are only small differences between the Unit Three content of these two courses, the main difference being that specific examples mentioned in the Australian Curriculum are moved to the glossary of the West Australian syllabus. Hence this text is also suitable for anyone following Unit Three of the Australian Curriculum course, *General Mathematics*.

The book contains text, examples and exercises containing many carefully graded questions. A student who studies the appropriate text and relevant examples should make good progress with the exercise that follows.

The book commences with a section entitled **Preliminary work**. This section briefly outlines work of particular relevance to this unit that students should either already have some familiarity with from the mathematics studied in earlier years, or for which the brief outline included in the section may be sufficient to bring the understanding of the concept up to the necessary level.

As students progress through the book they will encounter questions involving this Preliminary work in the **Miscellaneous exercises** that feature at the end of each chapter.

These miscellaneous exercises also include questions involving work from preceding chapters to encourage the continual revision needed throughout the unit.

Some chapters commence with, or contain, a '**Situation**' or two for students to consider, either individually or as a group. In this way students are encouraged to think and discuss a situation, which they are able to tackle using their existing knowledge, but which acts as a forerunner and stimulus for the ideas that follow. Students should be encouraged to discuss their solutions and answers to these situations and perhaps to present their method of solution to others. For this reason answers to these situations are generally not included in the book.

At times in this series of books I have found it appropriate to go a little beyond the confines of the syllabus for the unit involved. In this regard, in the chapter introducing recursively-defined sequences I sometimes go beyond the types specifically mentioned in the syllabus, when considering networks I include use of the word traversable and the idea of a tree and when considering linear regression I include mention of the 'x on y' regression line. I leave it up to the readers and teachers to decide whether to cover these aspects or not.

Alan Sadler

CONTENTS

IMPORTANT NOTE

This series of texts has been written based on my interpretation of the appropriate *Mathematics Applications* syllabus documents as they stand at the time of writing. It is likely that as time progresses some points of interpretation will become clarified and perhaps even some changes could be made to the original syllabus. I urge teachers of the *Mathematics Applications* course, and students following the course, to check with the appropriate curriculum authority to make themselves aware of the latest version of the syllabus current at the time they are studying the course.

Acknowledgements

As with all of my previous books I am again indebted to my wife, Rosemary, for her assistance, encouragement and help at every stage.

To my three beautiful daughters, Rosalyn, Jennifer and Donelle, thank you for the continued understanding you show when I am 'still doing sums' and for the love and belief you show.

In particular I thank Rosalyn for her input and I thank my good friend and ex-colleague Theo Wieman for his much appreciated wise counsel.

To the delightfully supportive team at Cengage, I thank you all.

Alan Sadler

PRELIMINARY WORK

This book assumes that you are already familiar with a number of mathematical ideas from your mathematical studies in earlier years.

This section outlines the ideas which are of particular relevance to Unit Three of the *Mathematics Applications* course and for which some familiarity will be assumed, or for which the brief explanation given here may be sufficient to bring your understanding of the concept up to the necessary level.

Read this 'Preliminary work' section and if anything is not familiar to you, and you don't understand the brief explanation given here, you may need to do some further reading to bring your understanding of those concepts up to an appropriate level for this unit. (If you do understand the work but feel somewhat 'rusty' with regards to applying the ideas some of the chapters afford further opportunities for revision as do some of the questions in the miscellaneous exercises at the end of chapters.)

- Chapters in this book will continue some of the topics from this Preliminary work by building on the assumed familiarity with the work.

- The miscellaneous exercises that feature at the end of each chapter may include questions requiring an understanding of the topics briefly explained here.

Number

The understanding and appropriate use of the rule of order, fractions, decimals, percentages, rounding, truncation, square roots and cube roots, numbers expressed with positive integer powers, e.g. 2^3, 5^2, 2^5 etc, expressing numbers in standard form, e.g. $2.3 \times 10^4 \ (= 23\,000)$, $5.43 \times 10^{-7} \ (= 0.000\,000\,543)$, also called scientific notation, and familiarity with the symbols $>$, $\geq$, $<$ and $\leq$ is assumed.

Percentages

It is assumed that you are able to express one quantity as a percentage of another quantity.

For example: 21 out of 50 is 42% $\dfrac{21}{50} \times 100$

35 out of 78 is 44.87%, correct to two decimal places. $\dfrac{35}{78} \times 100$

197 out of 5063 is 4%, to the nearest whole percent. $\dfrac{197}{5063} \times 100$

It is also assumed that you are able to increase or decrease an amount by a certain percentage by appropriate multiplication.

For example: To increase an amount by 25% we multiply by 1.25.
To decrease an amount by 5% we multiply by 0.95.

Data

Data collection is often carried out to investigate some aspect of our lives. The methods by which we collect that data can vary as can the types of data we collect. For some investigations we collect the data ourselves by asking questions, by measuring, by experiment, etc. This is called **primary data** – data we have collected ourselves.

Sometimes it is appropriate to use the data already collected by others. For us this would be **secondary data** – data collected by others.

If we ask someone what their favourite colour is or how tall they are the answers we get will vary from one person to another. Not everyone has blue as their favourite colour, not everyone is 163 cm tall, etc. The responses will vary. Favourite colour and height are examples of **variables**.

If we are considering *one* variable, for example favourite colour, any data we collect will be **univariate**. However if we wanted to investigate how favourite colour may change with the age of a person we would be considering two variables – favourite colour and age. In such cases we would be considering **bivariate** data.

Consider the following questions that might be asked in some data collection activities:

> How do you get to school (or work) – walk, cycle, car, bus, train, other?
> Are you male or female?
> What is your favourite colour?
> How would you rank your fitness level – low, medium, high?
> What is the number of the house you live in?

Each of these questions allow us to place the person who responds into a particular **category** or **group**. For example the group of people who walk to school, or the group of males, or the group of people whose favourite colour is red, or the group of people who consider their fitness level to be high, or the group of people who live in even numbered houses, etc. The variable concerned, be it mode of transport, gender, favourite colour, fitness level or house number are all examples of **categorical variables**. Data associated with a categorical variable is called **categorical data**.

Notice that the first three questions each involve categories for which order is irrelevant. In general it would be pointless to suggest an order to the modes of transport, or to the gender categories, or place red ahead of blue as a category. We may rank order according to the numbers of people in that category but not on the basis of the category names themselves which have no natural order about them. Such categorical variables are called **nominal categorical variables** (name – nominal).

However the last two questions involve categories that do have a natural order about them. These questions involve **ordinal categorical variables**, i.e. a categorical variable that does have some natural order about it (order – ordinal).

ISBN 9780170394994

Notice though that whilst categorical variables can have categories that have numbers assigned to them these numbers are simply labels, they have no numerical significance. Houses numbered 11, for example, are not necessarily bigger, better or more expensive than houses numbered 10. The house number is simply a label.

Categorical variables

Nominal
Based purely on the names of the various categories, no order is suggested.

Ordinal
The names of the categories, even though only labels, do suggest an order.

Now consider the following questions that might be asked in some data collection activities:

How many people live in your house?
How many people in your Mathematics class?
How tall are you?
How far have you walked today?

The responses to each of these questions will be **numerical** and the number will not just be a label, it will indicate a size or an amount. Some form of counting or measuring will be required to be done, or to have been done, for the response to be given. In each case the variable concerned, be it the number of people living in your house, the number of people in your Mathematics class, how tall you are, or how far you have walked today, are all examples of **numerical variables**. Data associated with a numerical variable is called **numerical data**.

Notice that the first two questions each involve responses that can only take specific values, in this case integer values. We cannot have 2.3 people living in a house, we cannot have 28.6 people in a Mathematics class. Numerical variables that take specific values taken from a finite set of values, usually the counting numbers 0, 1, 2, 3, … , are called **discrete variables**. Data associated with discrete variables is **discrete data**.

However the last two questions each involve responses that can take any value (usually within some realistic range). We can be 164.5 cm tall, we can walk 5.278 km in a day. Responses no longer have to be integer values and can now be taken from an infinite set of possible values. In practice the limit on the values taken are those of reasonableness (for example, we cannot have 18.24 metres for the height of someone) and the accuracy of the measurement instrument used. Numerical variables that can take any value in an interval are called **continuous variables.** Data associated with a continuous variable is **continuous data**.

Generally, if counting is involved we have a discrete variable, if measurement is involved we have a continuous variable.

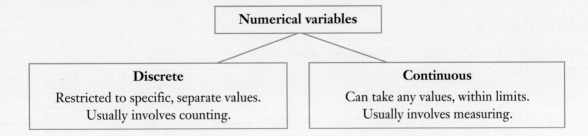

Numerical variables

Discrete
Restricted to specific, separate values.
Usually involves counting.

Continuous
Can take any values, within limits.
Usually involves measuring.

ISBN 9780170394994

Measures of central tendency

It is assumed that you are familiar with using the **mean**, the **median** and the **mode** as measures that can be used to summarise a set of data.

The mean and the median each indicate a 'central score'. The mode is often included in these 'averages' but there is no guarantee that the mode is any 'central' measure.

Measures of dispersion

It is assumed that you are familiar with using the **range** and the **standard deviation** to describe the *spread* of data.

The range of a set of scores is the difference between the highest score and the lowest score and gives a simple measure of how widely the scores are spread. Whilst the range is easy to calculate it is determined using just two of the scores. For this reason it is of limited use.

The standard deviation is more representative of spread because it is obtained using all of the scores in a set, not just the highest and lowest

The standard deviation is found by finding how much each of the scores differs from the mean, squaring these values, finding the average of the squared values and then taking the square root of this average.

Many calculators can determine the standard deviation of a set of scores.

The statistical investigation process

Collecting data is one part of the *statistical investigation process*, a process that can be described by the following steps:

- Clarify the problem and formulate one or more questions that can be answered with data.
- Design and implement a plan to collect and obtain appropriate data.
- Select and apply appropriate graphical or numerical techniques to analyse the data.
- Interpret the results of this analysis and relate the interpretation to the original question and communicate findings in a systematic and concise manner.

(As stated in the Australian Curriculum Glossary for General Mathematics and the West Australian Glossary for Mathematics Applications.)

Graphical display of data

It is assumed that you are familiar with using boxplots and histograms to compare data sets.

Coordinates

It is assumed that you are familiar with the idea that points on a graph can be located by stating the coordinates of the point.

For example: point A has coordinates (3, 2),
point B has coordinates (−2, 3),
point C has coordinates (−3, −2),
point D has coordinates (2, −3).

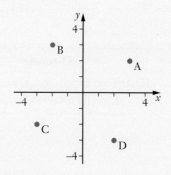

Straight line graphs

If points lie in a straight line their coordinates, (x, y), obey a rule of the form $y = mx + c$ and the relationship between x and y is said to be *linear*.

In this rule m is the gradient of the straight line and the point $(0, c)$ is where the line cuts the vertical axis.

Hence the line $y = 2x + 1$ has a gradient of 2 and cuts the y-axis at $(0, 1)$.

The line $y = -x + 5$ has gradient of -1 and cuts the y-axis at $(0, 5)$.

In the equation $y = mx + c$ the value of m, the gradient, tells us the amount by which y increases for each unit increase in x.

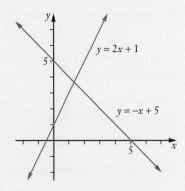

Solving equations

It is assumed that you are already familiar with the idea that solving an equation involves finding the value(s) the unknown can take that make the equation true.

For example, $x = 5.5$ is the solution to the equation $\quad 15 - 2x = 4$
because $\qquad\qquad\qquad\qquad\qquad\qquad\qquad 15 - 2(5.5) = 4.$

To solve an equation, e.g. $15 - 2x = 4$, we could proceed *mentally*:

$$\text{We know that fifteen take } eleven \text{ equals four.}$$
$$\text{Thus } 2x = 11 \text{ and so } x = 5.5.$$

We could use the *solve facility* of some calculators.

Or we could use a *step by step approach* to isolate x:

Given the equation:	$15 - 2x = 4$
Add $2x$ to both sides to make the x term positive:	$15 = 4 + 2x$
Subtract 4 from both sides to isolate $2x$:	$15 - 4 = 2x$
$\therefore$	$11 = 2x$
Divide both sides by 2 to isolate x:	$5.5 = x$
$\therefore$	$x = 5.5$

It is also anticipated that you are able to solve more involved linear equations, for example:

$$5x - 7 - 2x - 8 = 30 - 6x \qquad \text{Solution:} \quad x = 5$$

$$\frac{x + 3}{2} - 1 = 5 \qquad \text{Solution:} \quad x = 9$$

$$2(x + 3) - 3(2x + 1) = -5 \qquad \text{Solution:} \quad x = 2$$

and simultaneous linear equations, for example:

$$\begin{cases} 3x + 2y = 11 \\ x + 2y = 1 \end{cases} \qquad \begin{cases} 5x - 2y = 6 \\ 3x + 2y = 26 \end{cases} \qquad \begin{cases} 2x + 3y = 12 \\ x + 4y = 11 \end{cases}$$

Solution: $x = 5, y = -2$ $\qquad$ Solution: $x = 4, y = 7$ $\qquad$ Solution: $x = 3, y = 2$

ISBN 9780170394994

Matrices

It is assumed that you are familiar with the rows and columns, *rectangular array*, form of presenting information called a **matrix**, an example of which is shown on the right.

$$\begin{bmatrix} 5 & 2 & 1 & 2 & 10 & 5 \\ 5 & 2 & 1 & 2 & 4 & 5 \\ 5 & 2 & 0 & 3 & 7 & 6 \\ 5 & 2 & 0 & 3 & 4 & 11 \\ 5 & 3 & 2 & 0 & 8 & 2 \\ 5 & 2 & 0 & 3 & 5 & 9 \end{bmatrix}$$

This matrix has 6 rows and 6 columns. We say it is a six by six matrix, (written 6×6). This notation gives the **size** or **dimensions** of the matrix.

Matrices do not have to have the same number of rows as they have columns, however those that do are called **square matrices**.

The matrix on the right is a 6×6 square matrix.

Some matrices of other sizes are shown below.

$$\begin{bmatrix} 1 & 0 & 4 \\ 3 & 2 & 0.5 \end{bmatrix} \qquad \begin{bmatrix} 2 & 5 \\ 11 & -2 \end{bmatrix} \qquad \begin{bmatrix} 3 \\ 2 \\ -1 \end{bmatrix} \qquad \begin{bmatrix} 1 & 0 & 2 & 3 \\ 1 & 2 & -1 & 0 \\ 0 & 1 & 0 & 3 \end{bmatrix}$$

2 rows and 3 columns.   2 rows and 2 columns.   3 rows and 1 column.   3 rows and 4 columns.
A 2×3 matrix.   A 2×2 matrix.   A 3×1 matrix.   A 3×4 matrix.
(A square matrix)

A matrix consisting of just one column, like the third matrix above, is called a **column matrix**.

Any matrix consisting of just one row is called a **row matrix**. For example:

$$\begin{bmatrix} 5 & 0 & -2 & 1 \end{bmatrix}$$

A square matrix having zeros in all spaces that are not on the **leading diagonal** is called a **diagonal matrix**.

$$\begin{bmatrix} 5 & 0 & 0 & 0 \\ 0 & 0 & 0 & 0 \\ 0 & 0 & 2 & 0 \\ 0 & 0 & 0 & -4 \end{bmatrix}$$

We commonly use capital letters to label different matrices. The corresponding lower case letters, with subscripted numbers, are then used to indicate the row and column a particular entry or **element** occupies.

For the matrix A shown on the right the element occupying the 3rd row and 2nd column is the number 7. Thus $a_{32} = 7$.

Similarly

$a_{11} = 2,$
$a_{12} = 0,$
$a_{13} = -1,$ etc.

$$A = \begin{bmatrix} 2 & 0 & -1 & 3 \\ 1 & 6 & -3 & 9 \\ 5 & 7 & 8 & 4 \end{bmatrix}$$

ISBN 9780170394994

To add (or subtract) matrices we add (or subtract) the elements occurring in corresponding locations. To do this there must be elements in corresponding locations. Thus we can only add or subtract matrices that are the same size as each other. For example:

if $A = \begin{bmatrix} 1 & 0 & 2 & 3 \\ 4 & -2 & 3 & 5 \\ 2 & 1 & -3 & 4 \end{bmatrix}$ and $B = \begin{bmatrix} 2 & 1 & -3 & 2 \\ 5 & 1 & 2 & 4 \\ 3 & 2 & 0 & -5 \end{bmatrix}$ then $A + B = \begin{bmatrix} 3 & 1 & -1 & 5 \\ 9 & -1 & 5 & 9 \\ 5 & 3 & -3 & -1 \end{bmatrix}$

and similarly $A - B = \begin{bmatrix} -1 & -1 & 5 & 1 \\ -1 & -3 & 1 & 1 \\ -1 & -1 & -3 & 9 \end{bmatrix}$

To multiply a matrix by a number we multiply each element of the matrix by that number. (This is referred to as 'multiplication by a **scalar**'.)

Equal matrices must be the same size and have all corresponding elements equal.

Thus if $\begin{bmatrix} a & b & c \\ d & e & f \end{bmatrix} = \begin{bmatrix} 2 & 3 & -5 \\ 1 & 0 & -2 \end{bmatrix}$ then $\begin{matrix} a = 2 & b = 3 & c = -5 \\ d = 1 & e = 0 & f = -2 \end{matrix}$

Matrices can be multiplied together if the number of columns in the first matrix equals the number of rows in the second matrix.

To multiply two matrices we 'spin' each row of the first matrix to align with each column of the second, multiply and then add:

If $A = \begin{bmatrix} 2 & 1 & 3 \\ 0 & -1 & 2 \end{bmatrix}$ and $B = \begin{bmatrix} 1 & 2 \\ 4 & -1 \\ 1 & -3 \end{bmatrix}$ the product AB is found as shown below.

$$\begin{bmatrix} 2 & 1 & 3 \\ 0 & -1 & 2 \end{bmatrix}\begin{bmatrix} 1 & 2 \\ 4 & -1 \\ 1 & -3 \end{bmatrix} = \begin{bmatrix} (2)(1)+(1)(4)+(3)(1) & (2)(2)+(1)(-1)+(3)(-3) \\ (0)(1)+(-1)(4)+(2)(1) & (0)(2)+(-1)(-1)+(2)(-3) \end{bmatrix}$$

$$= \begin{bmatrix} 9 & -6 \\ -2 & -5 \end{bmatrix}$$

Many calculators will accept data in matrix form and can then manipulate these matrices in various ways.

$\begin{bmatrix} 1 & 2 & 4 \\ 0 & -4 & 5 \end{bmatrix} + \begin{bmatrix} 3 & 5 & -2 \\ 1 & 0 & -2 \end{bmatrix}$

$\begin{bmatrix} 4 & 7 & 2 \\ 1 & -4 & 3 \end{bmatrix}$

$\begin{bmatrix} 2 & 1 & 3 \\ 0 & -1 & 2 \end{bmatrix} \times \begin{bmatrix} 2 & 1 \\ 4 & -1 \\ 1 & -3 \end{bmatrix}$

$\begin{bmatrix} 9 & -6 \\ -2 & -5 \end{bmatrix}$

Note: In the product AB we say that B is *premultiplied* by A
or that A is *postmultiplied* by B.

Networks and matrices

If you have previously completed Unit One of *Mathematics Applications* you will be familiar with

    a   the idea of road systems and social interactions being displayed in the diagrammatic form shown below (called networks),

and   **b**   the associated matrix for such systems.

For example:

A road network	**A social interaction network**
The network shows road links between towns A, B, C and D.	The network shows who in a group of five people, A, B, C, D and E has visited the house of somebody else in the group.

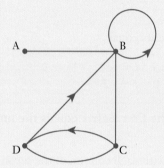 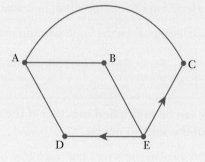

The associated matrix:

$$
\text{From}\quad
\begin{array}{c}
\\ A \\ B \\ C \\ D
\end{array}
\begin{array}{cccc}
A & B & C & D \\
\left[\begin{array}{cccc}
0 & 1 & 0 & 0 \\
1 & 1 & 1 & 0 \\
0 & 1 & 0 & 2 \\
0 & 1 & 1 & 0
\end{array}\right]
\end{array}
\qquad
\begin{array}{c}
\\ A \\ B \\ C \\ D \\ E
\end{array}
\begin{array}{ccccc}
A & B & C & D & E \\
\left[\begin{array}{ccccc}
0 & 1 & 1 & 1 & 0 \\
1 & 0 & 0 & 0 & 1 \\
1 & 0 & 0 & 0 & 0 \\
1 & 0 & 0 & 0 & 0 \\
0 & 1 & 1 & 1 & 0
\end{array}\right]
\end{array}
$$

Notice that in the social interaction matrix:

- We use a '1' to indicate 'has visited' and a '0' to indicate 'has not visited'.

- We use the convention that the matrix shows 'what the row person does to the column person'. For example the entry in row 5 (person E) and column 4 (person D) shows that person E has been to the house of person D.

- The 'has visited your own house' is really rather pointless information so we simply place zeros on the leading diagonal. On some occasions like this we might simply decide to put dashes on the leading diagonal. (However if we later wanted to multiply the matrix by a scalar or another matrix these dashes could cause a problem.)

ISBN 9780170394994

1.

Bivariate data

Situation

A statistician was asked to investigate whether countries with higher levels of consumption of a particular foodstuff tended to have higher death rates due to heart disease. Searching databases available from various groups concerned with international health the statistician compiled the following table:

	CONSUMPTION OF THE FOODSTUFF Kilograms per person per year	DEATHS DUE TO HEART DISEASE Per year per 100 000 of population
Country A	57	299
Country B	20	124
Country C	13	154
Country D	38	156
Country E	51	152
Country F	48	295
Country G	58	264
Country H	35	208
Country I	29	250
Country J	8	84
Country K	28	158
Country L	21	175
Country M	10	120
Country N	40	220
Country O	41	251
Country P	46	202
Country Q	53	248
Country R	64	290
Country S	66	296
Country T	47	268

What would be your conclusion?
(Justify your decision.)

The situation on the previous page asked you to consider whether the given data indicated that countries with higher levels of consumption of a particular foodstuff tended to have higher death rates due to heart disease.

The situation involves **two** variables: Consumption of a particular foodstuff,
and Death rates due to heart disease.

We now have *bivariate data*. Hence the title of this chapter.

With bivariate data we are often interested in investigating whether there is some relationship or association between the two variables. This was the case with the previous situation in which we were interested in whether high consumption of the particular foodstuff might explain higher death rates due to heart disease.

The two variables in the previous situation each involved *numerical* data. Did you consider plotting the bivariate data as a graph with one variable on one axis, the other on the other axis and the data for each country represented by a point on your graph? Such a *scattergraph* allows the *scatter* of points to be considered.

We will consider scattergraphs for bivariate data again later in this chapter but, for now, let us turn our attention to bivariate data for which one or both variables involve *categorical* data. Such data may involve categories such as

Male–Female,
Low–Medium–High,
States and Territories of Australia, etc.

For such data the more appropriate form of display is a two-way frequency table, rather than a scattergraph, but how would we investigate possible associations between two categorical variables from a frequency table?

Investigating two-way frequency tables

Two-way tables

Let us suppose that in a particular country the number of male and female Members of Parliament (MPs) from the two major political parties, A and B, are as shown in the table on the right.

	Party A	Party B
Male	215	104
Female	42	61

Including the various totals gives:

Various observations could be made from these tables, for example:

	Party A	Party B	Totals
Male	215	104	319
Female	42	61	103
Totals	257	165	422

- Party A and Party B each have more male MPs than female MPs.

- Whilst Party B has fewer MPs than Party A (165 compared with 257) it has more female MPs (61 compared with 42).

For some observations it may be clearer if the various entries are given as *proportions* of the total number (422), or perhaps as *percentages* of this total. Such tables are shown at the top of the next page with the one on the left showing proportions of the total number of 422 (given correct to two decimal places) and the one on the right showing percentages of the total (to the nearest whole percentage).

	Party A	Party B	Totals
Male	0.51	0.25	0.76
Female	0.10	0.14	0.24
Totals	0.61	0.39	1.00

	Party A	Party B	Totals
Male	51%	25%	76%
Female	10%	14%	24%
Totals	61%	39%	100%

Now observations like
- 51% of all of the MPs from the two parties are males from Party A,
- 76% of all the MPs from these two parties are male,

more readily come to mind.

Row and column percentages

We could also consider percentages of the row totals, called row percentages, as shown below left, and percentages of column totals, called column percentages, as shown below right. (Check that you agree with the various entries in each table.)

	Party A	Party B	Totals
Male	67%	33%	100%
Female	41%	59%	100%

	Party A	Party B
Male	84%	63%
Female	16%	37%
Totals	100%	100%

Now observations like
- 67% of the male MPs from these two parties are from Party A,
- 59% of the female MPs from these two parties are from Party B,
- 37% of Party B MPs are female compared with 16% of Party A MPs,

are more apparent.

Displaying the percentages from the table above right, as two proportionally divided columns, gives the graph shown on the right (sometimes referred to as a *stacked 100% column graph*). We can see from this diagram, as indeed we could from the table, that as we move from the Party A column to the Party B column the male-female proportions change noticeably. The proportion of MPs who are female is much greater in Party B than in Party A. I.e., with regard to gender balance it does matter which party we are considering. This suggests there is an *association* between the two categorical variables of Party and Gender of MP. If there were no association we would not expect the proportions to change so noticeably as we move from the Party A column of our graph to the Party B column.

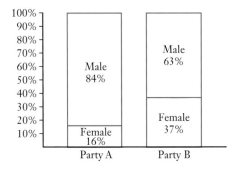

Hint: For ease of dividing each column into percentages, when drawing the column graph by hand make the column height 100 mm.

Choosing the Australian Federal Government, or one of the state or territory governments, research the number of male and female MPs in the major parties and present two-way tables, proportional tables and percentage tables, similar to those shown here, for your data.

Make some comments based on your findings.

1. Bivariate data ●●○○○○

Suppose instead that we investigated the mode of transport to work (bus, car, train, other) and gender. Further suppose that survey results gave rise to the following table:

	Bus	Car	Train	Other	Totals
Male	26	47	32	19	124
Female	35	70	45	24	174

This table gives rise to the table below showing row percentages, given to the nearest whole percentage. (Check that you agree with the percentages.)

	Bus	Car	Train	Other	Totals
Male	21%	38%	26%	15%	100%
Female	20%	40%	26%	14%	100%

Notice that the male-female proportions are very similar in each column, suggesting the lack of any association between gender and mode of transport to work. I.e., with regard to mode of transport to work it does not matter what gender the person is.

The proportional column graph shown on the right similarly indicates that it does not matter whether we are considering male or female, the proportions using the various modes of transport are almost identical as we move from one gender column of the graph to the other.

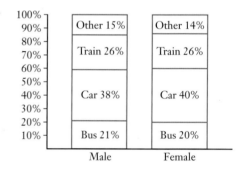

The proportions of the various modes of transport are not influenced by what gender we are considering. Contrast this with the gender balance of MPs considered earlier. The proportion of female MPs was certainly influenced by which party we were considering.

If in changing the category of one variable (e.g. changing from Party A to Party B or from male to female) the proportions in the various categories of the other variable change, then this suggests that the two variables are **associated**. Of course, we would expect some small changes to occur just because of the variation in the samples involved, so for the existence of an association to be strongly suggested we would need the changes in proportions to be significant.

In the MPs situation we looked at whether considering a different party meant that the proportion of male MPs to female MPs changed. To do this we split the party numbers up into male and female percentages.

In the modes of transport situation we looked at whether considering a different gender meant that the proportion of modes of transport changed. To do this we split the male and female numbers up into types of transport percentages.

As well as being able to construct and examine appropriate two-way tables and graphs to see if association is suggested, we should also be able to describe an association by commenting on the nature and magnitude of any changes in proportions that exist. The next two examples, and the exercise that follows, cover these aspects.

EXAMPLE 1

A number of men were asked which was their most preferred type of movie out of:

Crime, Romance or Horror.

The data obtained was sorted according to the age of the respondent and gave rise to the table shown below.

		Film type		
		Crime	**Romance**	**Horror**
Age in years	21 to 35	26	11	67
	36 to 50	35	9	29
	51 to 65	22	5	14

a Draw the frequency table showing row percentages (i.e. % based on row totals).

b Draw a proportional column graph involving three columns of equal height with one column for each of the three age groups.

c Comment on whether any association between the variables is suggested and write some comments describing the association.

Solution

a

		Film type			
		Crime	**Romance**	**Horror**	**Totals**
Age in years	21 to 35	25%	11%	64%	100%
	36 to 50	48%	12%	40%	100%
	51 to 65	54%	12%	34%	100%

b

c The ages of 36 to 50 and 51 to 65 show similar proportions of preferred film type but there is a clear difference in the proportions in the first column to those in the other two, suggesting an association between *age* and *preferred film type*.

The percentage of each age category preferring romance films is virtually unchanged at about 12%, irrespective of what age category we are considering.

Whereas about two-thirds of 21 to 35 year old males prefer horror movies only one-third of 51 to 65 year old males have horror movies as their preferred type.

Having crime as the preferred choice goes the other way with just one-quarter of 21 to 35 year old males having it as their preferred type whereas just over half of 51 to 65 year olds choose it as their preferred type of movie.

In part **c** of the previous example the comments given, after the first, are given by way of example. There are other correct comments that could be given.

EXAMPLE 2

Classifying people as one of the three categories:

Left handed, Right handed, Mixed/Ambidextrous,

a survey considering whether the 'handedness' of a firstborn child could be explained by the 'handedness' of the mother, produced the following frequency table.

		Handedness of mother		
		Left handed	Right handed	Mixed/Ambi
Handedness of firstborn child	Left handed	4	21	1
	Right handed	21	155	3
	Mixed/Ambi	2	8	1

Redraw the above table showing column percentages and use it to identify and comment on any patterns shown and whether any association between the handedness of a child and the handedness of the mother is suggested.

Solution

The table with column percentages (nearest percent) will be as shown below.

		Handedness of mother		
		Left handed	Right handed	Mixed/Ambi
Handedness of firstborn child	Left handed	15%	11%	20%
	Right handed	78%	84%	60%
	Mixed/Ambi	7%	4%	20%
	Totals	100%	99%*	100%

*The total of 99%, rather than 100%, is simply due to rounding.

Moving across the columns in the above table does show some variation of proportions, the most noticeable difference being in that of the proportions in the 'Mixed/Ambi' column compared to the others. However the percentages for that column are based on a very small total (5) and a change of just 1 of these 5 from one category to another would greatly change the percentages. Hence this final column is of limited use.

Considering the changes in the proportions in the first two columns the changes are not huge and could simply be due to the sample of people involved. Certainly a greater percentage of the right handed mothers had a right handed firstborn child (84%) than did left handed mothers (78%) but the difference in these percentages is only 6%. Similarly a higher percentage of left handed mothers had left handed firstborns (15%) than did right handed mothers (11%).

Thus the table suggests that whilst the handedness of the mother and the handedness of the firstborn could, perhaps, to some limited extent, be associated with each other the association is not particularly strong (and there were only 4 left handed firstborns of left handed mothers involved in the survey).

ISBN 9780170394994

Interestingly, the table above makes no mention of the handedness of the child's father. If we wanted to investigate whether there is some association between a child's handedness and that of the parents this aspect could be an important consideration.

Note: In the previous example the proportional column graph, though not requested, could also be drawn to assist interpretation if so desired.

Note

Tables given in this chapter have been created for the purposes of this book. They are not necessarily taken from real data but do attempt to reflect a situation that could exist.

Exercise 1A

1 Suppose that a survey asking a number of people their age and which they most preferred as a main course at a restaurant, out of beef, chicken, fish or other, gave rise to the two-way table shown below.

	Beef	Chicken	Fish	Other
18–30	10	7	3	1
31–43	9	10	12	4
44–56	7	17	15	3
Over 56	2	5	8	2

a How many people responded to the survey?

b How many of the respondents chose beef as their most preferred main course at a restaurant?

c What percentage of those choosing beef as their most preferred course at a restaurant were aged 31–43?

d What percentage of those aged 31–43 chose beef as their preferred main course at a restaurant?

2 Suppose that a survey investigating the gender and current marital situation of a number of adults gave rise to the two-way frequency table shown below.

For current marital situation respondees indicated *one* of the four categories:

single (and never married), married,
divorced (and not currently married), widowed (widow or widower).

	Single	Married	Divorced	Widowed
Male	49	68	24	21
Female	60	92	30	32

a How many people does the frequency table involve?

b How many males does the frequency table involve?

c What percentage of the males involved in the survey were divorced?

d What percentage of the divorced people involved in the survey were males?

e According to the survey results which is the greater, the percentage of males who are single or the percentage of females who are single?

3

	A	B	C	D
E	98	116	154	336
F	52	66	64	18

The two-way frequency table shown above involves two variables, one having categories A, B, C and D and the other having categories E and F.

Create 3 tables, together with appropriate totals (and percentages to nearest 1%):

a A table showing each entry as a percentage of the total (904).

b A table showing row percentages. (Percentages based on row totals.)

c A table showing column percentages. (Percentages based on column totals.)

4 Suppose that a survey of the preferred football code, out of Australian Rules, Rugby League, Rugby Union and Soccer and the state where the respondent lives, was conducted for a sample of 291 people living in four states. The results obtained gave rise to the following table:

	NSW	Queensland	Victoria	WA
Australian Rules	12	15	54	39
Rugby League	50	32	4	2
Rugby Union	20	13	6	6
Soccer	15	8	8	7

a Redraw the above table showing column percentages (to the nearest 1%).

b Draw a proportional column graph involving four columns of equal height with one column for each of NSW, Queensland, Victoria and WA.

c Make some comments summarising some of the information shown in the percentages table and column graph and in particular comment on whether any association between the preferred football code and the state in which a person lives is suggested.

5 In addition to the usual drug treatment for a particular disorder a medical facility also invites sufferers to follow one of three other programs, A, B or C. The success or failure of 119 people treated for the disorder gave rise to the following table.

	Success	Failure
Drug program only	18	10
Drug program + program A	13	5
Drug program + program B	24	8
Drug program + program C	29	12

a Redraw the above table showing row percentages (to the nearest 1%).

b Draw a proportional column graph involving four columns of equal height with one column for each of the four 'program categories'.

c Make some comments summarising some of the information shown in the percentages table and column graph and in particular comment on whether any association is suggested between success and failure and the programs.

6 The table below shows the numbers of males and females employed by a particular organisation, grouped according to age.

	30 and under	31 to 40	41 to 50	51 to 60	61 and over
Male	47	39	62	76	15
Female	51	34	58	35	6

a Redraw the above table showing column percentages (to the nearest 1%).

b Investigate how the proportions of males and females change as we move across the age categories. Is any association suggested? Comment on your findings and write some summary statements describing any differences you observe in gender percentages across the various age categories.

(The organisation had, in recent years, introduced a gender equality policy.)

7 A number of newly wedded couples, who did not yet have children, were asked:

If you were to have a daughter, which activity out of

Music,   Dance,   A Ball Sport,   Gymnastics,   Horse Riding

would you most like them to excel in?

The responses from 177 males and 194 females gave rise to the following table:

	Music	Dance	A ball sport	Gymnastics	Horse riding
Male	62	18	53	32	12
Female	73	39	27	39	16

a Redraw the above table showing row percentages (to the nearest 1%).

b Investigate how the proportions in each category of activity changes as we move across the gender categories. Comment on your findings and write some summary statements describing any differences you observe in activity percentages as we move across the gender categories.

8 A survey of 900 people aged 15 and over, and in full time employment, classified the employment of each person as being one of three types:

An employee
An independent, working under contract
A business owner selling goods to the public

The survey gave rise to the following table:

	15 to 29	30 to 44	45 to 59	60 and over
Employee	203	183	231	99
Independent	7	17	38	24
Business owner	4	15	40	39

a Redraw the above table showing column percentages (to the nearest 1%).

b Investigate how the proportions in each employment category change as we move across the age categories. Comment on your findings and write some summary statements describing any differences you observe in the percentages in each employment type as we move across the age categories.

1. Bivariate data ●●●○○○

9 Suppose that a number of adults were asked to rate themselves on a happiness scale:

Usually very happy

Usually happy

Neutral

Usually sad

and that the results of the survey were as given in the table below.

	Usually very happy	Usually happy	Neutral	Usually sad
Male	23	34	20	9
Female	36	52	26	11

Redraw the above table showing row percentages (to the nearest 1%) and use it to identify and comment on any patterns shown and whether any association between happiness and gender is suggested.

Explanatory variable and response variable

The situation at the beginning of this chapter considered whether countries with higher levels of consumption of a particular foodstuff tended to have higher death rates due to heart disease. Researchers would be interested to investigate whether the higher death rates were perhaps in some way *explained* by the higher consumption of the foodstuff. In this situation we say that the consumption of the foodstuff is the **explanatory variable**, also called the independent variable or predictor variable.

Were the higher death rates due to heart disease in some way a *response* to the higher levels of consumption of the foodstuff? We say that the death rate is the **response variable**, also called the dependent variable or predicted variable.

> When investigating relationships in bivariate data, the **explanatory variable** is the variable used to explain or predict a difference in the **response variable**.

Earlier in this chapter we considered whether the balance of male/female MPs could depend upon the party the MP was in.

Explanatory variable: The party the MP was in.

Response variable: Whether the MP was male or female.

We looked at whether the choice of mode of transport to work could in some way be explained by gender.

Explanatory variable: Gender.

Response variable: Choice of mode of transport to work.

We also considered whether the preferred type of movie could be explained by age.

Explanatory variable: Age.

Response variable: Preferred type of movie.

We might consider whether high blood pressure increases the risk of a heart attack.

Explanatory variable: High blood pressure.

Response variable: Risk of heart attack.

If you find it difficult to identify which variable is the explanatory variable and which the response variable it can be helpful to ask:

Which variable explains differences in the other?

Which variable would be used to predict the other?

Which occurs first? As in high blood pressure (explanatory) and heart attack (response).

ISBN 9780170394994

If we know what it is we are trying to investigate the identification of explanatory variable and response variable usually makes sense. However, if all we are given are two categorical variables it may not be obvious which of the two variables is the explanatory variable and which is the response variable. Changes in one variable A may not give any explanation of the changes in variable B and variable B may not respond in any way to changes in variable A. Situations could occur where in most cases it would be logical to use variable A to predict variable B but just occasionally a situation could need us to predict variable B from variable A. However, in this chapter, if a question requires us to decide which variable is which such 'unusual' situations are avoided and the identification of which variable is the explanatory variable and which is the response variable will follow what should seem to be the logical choice.

Which to use – row percentages or column percentages?

In the MPs situation encountered earlier in this chapter we considered whether the party an MP was in would explain the male-female balance. The political parties A and B formed the explanatory variable and gender (male-female) the response variable. The table gave the political parties as columns and we viewed the table of column percentages (i.e. percentages based on column totals) and created our proportional column graph from these percentages.

	Party A	Party B
Male	84%	63%
Female	16%	37%
Totals	100%	100%

In the modes of transport to work situation we considered whether gender explained any variations in choice of mode of transport to work. Now gender is the explanatory variable and mode of transport the response variable (although we then went on to show no association indicating no response to gender). The table gave gender as the rows in the table and we viewed the table of row percentages, i.e. percentages based on row totals, and created our proportional column graph from these percentages.

	Bus	Car	Train	Other	Totals
Male	21%	38%	26%	15%	100%
Female	20%	40%	26%	14%	100%

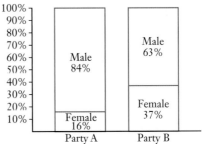

In each case the decision about which percentage table to use depended on which variable involves the categories we want to consider 'moving across'.

> We wanted to consider male-female proportions as we moved across political parties. Thus we wanted percentages of male-female within each political party.

> We wanted to consider mode of transport proportions as we moved across gender categories. Thus we wanted percentages of mode of transport within each gender.

For each situation, with thought, we could determine whether the row percentages or the column percentages were the more appropriate. However, if you prefer a rule:

- If the explanatory variable uses the rows to show its different categories use row percentages (percentages based on row totals).

- If the explanatory variable uses the columns to show its different categories use column percentages (percentages based on column totals).

I.e., in each case, split the category totals of the explanatory variable into percentages.

EXAMPLE 3

A number of people were asked whether they were in favour of increasing the number of driving hours a learner driver must do before gaining a licence, not in favour or were undecided on the issue. The results, by age of respondent, led to the following table.

	Age of respondent (in years)				
	18 to 29	30 to 41	42 to 53	54 to 65	Over 65
In favour of more hours	3	10	12	5	5
Not in favour of more hours	24	32	26	13	8
Undecided	3	8	10	8	9

These results are to be used to investigate whether the way a person feels about this issue could be explained by the age of the person.

a Which is the explanatory variable and which is the response variable?

b Draw the table showing either row or column percentages as appropriate and use your table to create a proportional column graph.

c Comment on whether there seems to be an association between the variables, explaining your reasoning and describing the association.

Solution

a We wish to consider whether the response can be explained by age. Age is the explanatory variable and the response to the question is the response variable.

b We wish to see if the response percentages change as we move across the age categories. Thus we need column percentages. Or, using the rule, the explanatory variable uses columns to display its categories so we want column percentages.

	Age of respondent (in years)				
	18 to 29	30 to 41	42 to 53	54 to 65	Over 65
In favour	10%	20%	25%	19%	23%
Not in favour	80%	64%	54%	50%	36%
Undecided	10%	16%	21%	31%	41%
Totals	100%	100%	100%	100%	100%

ISBN 9780170394994

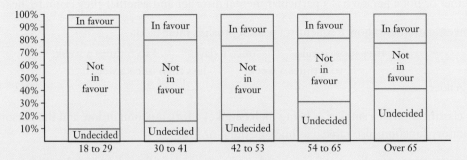

c The change in the proportions as we move across the age categories suggests that there is an association between the response and age.

Whilst in each age category the 'Not in favour' is much larger than the 'In favour' the size of the 'Not in favour' decreases as the age category increases.

The percentage undecided shows fairly steady increase as age category increases, from 10% of 18 to 29s undecided up to 41% of over 65s undecided.

The proportion 'In favour' is least in the 18 to 29 age category (10%) which is roughly half the proportion in each of the other age categories (from 19% to 25%).

Exercise 1B

For numbers 1 to 16 identify the explanatory variable and the response variable.

1 A survey investigating whether preferred film type can be explained by the gender of the person asked to give their preference.

2 A survey investigating whether the activity a parent would most like a daughter to excel at can be explained by the gender of the parent.

3 A survey investigated whether there was any evidence that the likelihood of a person developing respiratory problems in the teenage years could be associated with the age of the person's mother at the time the person was born.

4 A survey investigating whether the left or right handedness of a child was associated with the left or right handedness of the child's mother.

5 A survey investigating whether a person's preferred football code, out of AFL, Rugby League, Rugby Union and Soccer could be explained by the state a person lives in.

6 Investigating how the grade a student gets in an exam may relate to the amount of study a student does for that exam.

7 Investigating how the number of babies born in a country per year might relate to the population of the country.

8 Investigating into how well a person sleeps at night and a rating of the person's grumpiness the next morning.

9 Whether or not a person has a particular mental disorder and whether they claim to 'hear voices'.

10 An investigation into how frequently a person brushes their teeth:

after most meals,   once a day,   4 to 6 times a week,
1 to 3 times a week,   less than once a week.

and gender.

11 The classification of a smoker as being high (smokes a lot), medium or low and the person's respiratory function, also ranked as high, medium or low.

12 An investigation into the age of women when they have their first child and the age of each woman's mother when the mother had her first child.

13 Heating milk in a saucepan with the two variables being

the temperature of the milk
and   the time the heat has been applied for.

14 Lack of sleep and level of tiredness.

15 Stress levels and heart disease.

16 How dangerous a driver is to other road users and the amount of alcohol the driver has consumed recently.

17 Suppose a survey was carried out to investigate whether there was any evidence to suggest that the likelihood of someone developing respiratory problems in their teenage years was associated with the amount that their mother engaged in a particular activity during pregnancy. Further suppose that the survey produced the following results.

		Level of respiratory problems in teenage years		
		None	Moderate	Severe
Engaged in activity during pregnancy	Not at all	63	17	5
	Sometimes	18	16	7
	Often	10	18	6

a Which is the explanatory variable and which is the response variable in this situation?

b Are the explanatory variable's categories displayed using different rows or different columns?

c Recreate the table showing either row or column percentages as appropriate.

d Use your table to create a proportional column graph with equal height columns each representing 100% of an appropriate category.

e Comment on whether there seems to be an association between the variables, explaining your reasons and describing the association.

18 Could an individual's preferred soft drink be explained in any way by the age of the individual? Let us suppose that to investigate this question a group of 416 people, aged from 16 to 65, were asked to nominate their most preferred drink out of:

A fizzy drink	Fruit juice	Coffee
Tea	Milk/Flavoured milk	Water

The responses led to the creation of the following table of most preferred soft drink:

Age	Number of people nominating the drink as most preferred soft drink					
	Fizzy drink	Fruit juice	Coffee	Tea	Milk/ Flavoured milk	Water
16 to 25	28	20	7	3	13	9
26 to 35	28	20	22	12	14	21
36 to 45	11	10	14	8	5	17
46 to 55	8	14	23	15	6	27
56 to 65	5	10	15	11	4	16

a Which is the explanatory variable and which is the response variable in this situation?

b Recreate the table showing either row or column percentages as appropriate.

c Comment on whether there seems to be an association between the variables, explaining your reasons and describing the association.

19 In an attempt to determine what factors might explain the level of salary for a person working in a particular field, a nationwide survey of a number of people aged between 30 and 40 and in full-time employment in the particular field were asked, amongst other things, their salary, their blood group and the highest level of education they had achieved. With the salaries then grouped into the categories Medium, High and Very high, the results led to the following frequency tables.

		SALARY			
		Medium	High	Very high	Totals
BLOOD GROUP	A	85	53	12	150
	B	21	17	3	41
	AB	7	6	2	15
	O	106	67	17	190
	Totals	219	143	34	396

		HIGHEST LEVEL OF EDUCATION				
		School	Training diploma or certificate	Degree	Higher degree	Totals
Salary	Medium	40	52	111	16	219
	High	10	25	71	37	143
	Very high	2	4	16	12	34
	Totals	52	81	198	65	396

a Why do you think the survey was limited to people in full-time employment and to people of a particular age group?

b Use the first table to investigate whether blood group might explain the salary level of a person aged between 30 and 40 and working full-time in this particular field. State which variable is the explanatory variable, which is the response variable, present an appropriately percentaged version of the table, include a graph if you wish and comment on your findings.

c Use the second table to investigate whether level of education might explain the salary level of a person aged between 30 and 40 and working full time in this particular field. State which variable is the explanatory variable, which is the response variable, present an appropriately percentaged version of the table, include a graph if you wish and comment on your findings.

20 An organisation developed a program to help people give up smoking and wondered whether a smoker's success or failure on the program could in any way be explained by whether or not the smoker's partner was also a smoker and, if the partner was a smoker, whether they were also attempting to give up smoking.

The organisation contacted a number of people ('subjects') who had been on the program and who had partners. (To avoid double counting, if both subject and partner had been on the program only one of the two would feature as a subject.)

The subjects were asked if they felt that the program had been successful or not, if their partner was also an active smoker at the time and, if so, whether the partner was also trying to give up smoking.

From the responses of 107 people the following table was created.

		Subject's opinion of the program	
		A success	**Not a success**
Smoking status of partner	**Not a smoker**	30	20
	Smoker trying to give up	15	6
	Smoker not trying to give up	18	18

The organisation operated the stop smoking program in two cities, City A and City B, and wondered if the success rate differed from one city to the next. The organisers knew that of the 107 people who responded to the previous questions, 48 were from people who did the program in city A and, of these, 29 said they considered the program a success.

- Is a smoker's opinion of whether the program was a success or not associated with the smoking 'status' of their partner?

- Was the program in one city achieving greater success than in the other city?

- Can the organisers conclude that people attempting to give up smoking have a greater chance of success by being on the program than if they are not on the program?

Investigate. Comment. Write a report.

21 Depending on a person's sleep-wake patterns and preferred activity time throughout the day, they may be classified as a morning, afternoon or evening person. These categories indicate when the person feels most alert and performs their best at mental and physical tests. These time of day variations potentially have implications for society. For example, a morning type person, who would typically perform less well in the evening, may be well advised to avoid night duty.

Might the classification in any way be associated with gender? If so, perhaps workplaces that feature significantly more people of one gender than the other should start earlier, or perhaps later, in the day.

Might the classification in any way be associated with age? If so, perhaps the start time at schools should be adjusted accordingly. If teenagers were to be found to be morning people whilst younger children were found to be afternoon people should the respective school start times reflect this?

For the purposes of this question let us suppose that an investigation into such 'morningness/afternoon/eveningness' classification led to the following frequency tables, the entry in each cell of the tables indicating the number of people in that category.

	Morning	Afternoon	Evening	Totals
Teenagers	6	44	26	76
20 to 39	23	63	19	105
40 to 59	21	55	14	90
60 and over	23	35	7	65
Totals	73	197	66	336

	Morning	Afternoon	Evening	Totals
Female	42	112	35	189
Male	31	85	31	147
Totals	73	197	66	336

Investigate, comment and write a report based on this data.

Collect, organise and present data, and write a report, to consider one of the following questions (or a different question negotiated between you and your teacher).

- Is the male/female balance amongst the top achievers in a school influenced by the year group being considered?

- Is the mode of transport chosen to get to school explained to some extent by the age of the school student?

- Does the male/female balance of teachers within a subject area change much as we move across subject areas? E.g. English, Mathematics, Phys Ed, etc.

- Is there an association between opinion towards capital punishment (in favour of it / no opinion about it / against it) and gender?

Scattergraphs

Let us consider again the situation at the beginning of this chapter which involved an investigation into whether countries with higher levels of consumption of a particular foodstuff tended to have higher death rates due to heart disease. We were given foodstuff consumption figures (in kilograms per person per year) and the numbers of deaths due to heart disease (in deaths per year per 100 000 of population). Both of these variables, foodstuff consumption and deaths due to heart disease, involve *numerical* data, rather than the categorical data we have been considering more recently. Hence, rather than a two-way frequency table, a *scattergraph*, with one variable on one axis, the other on the other axis and the data for each country represented by a point on your graph, becomes the more suitable form of display. Such a scattergraph, shown below, allows the *scatter* of points to be considered.

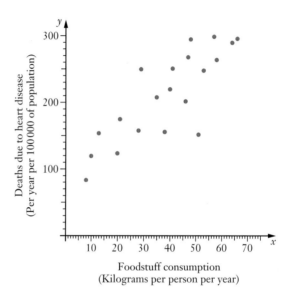

Foodstuff consumption
(Kilograms per person per year)

Notice that the points tend to form an oval shape from lower left to upper right, as shown on the right.

This indicates that in general, countries with a lower consumption of the foodstuff tend to have the lower death rate due to heart disease and those with the higher consumption of the foodstuff tend to have the higher death rate due to heart disease. There seems to be an **association**, or **correlation**, between the foodstuff consumption and the death rate due to heart disease.

Notice the use of phrases like *in general* and *tend to* in the previous paragraph.

The statement attempts to summarise an underlying pattern or *trend* suggested by the points. Not every individual will necessarily fit the trend, but we can comment on the general behaviour that the points suggest.

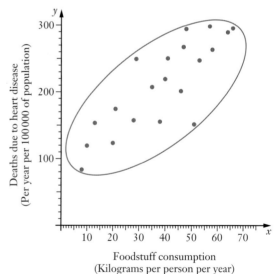

Foodstuff consumption
(Kilograms per person per year)

We could attempt to draw a straight line that seems to 'best fit' this data, as shown on the right.

In this way we fit a **linear model** to the situation.

We could then use this model to make predictions.

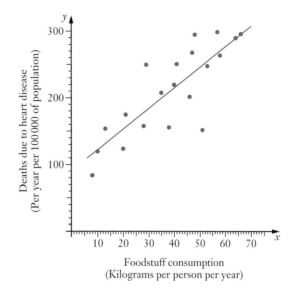

For example:

Using our **line of best fit** we could suggest that a country with an intake of 45 kilograms per person per year of the foodstuff, might have a heart disease death rate of approximately 230 persons per year per 100 000 of population (see the dotted orange lines in the scattergraph on the right).

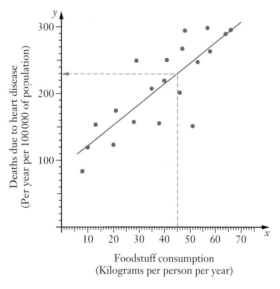

Note: In this foodstuff consumption / death rate due to heart disease situation we are investigating whether the death rate due to heart disease can be explained by the consumption of a particular foodstuff. I.e., is the death rate due to heart disease in any way associated with the amount of the foodstuff consumed? The foodstuff consumption rate is **the explanatory variable** (or **independent variable**), and the heart disease death rate the **response variable** (or **dependent variable**). We tend to plot the explanatory variable, or independent variable, on the horizontal axis and the response variable, or dependent variable, on the vertical axis.

If it is not clear which variable is which, then plot on the horizontal axis the data that is recorded first and plot on the vertical axis the data that is recorded second.

If we are investigating how something changes over *time* one of the variables will be *time* and we have what we call **time series data**. The time variable is the independent variable and is plotted on the horizontal axis. Time series data will be considered in greater detail in *Mathematics Applications Unit Four*.

Reliability of our predictions

When considering just how reliable our prediction might be, the things we need to consider are:

- Does the distribution of points on the scattergraph suggest that using a linear model, or perhaps some other model, is a good idea? For example, three scattergraphs are shown below. In one the distribution of points suggests a linear model would be appropriate, in one the distribution of points could be sensibly summarised using a curve and in one the distribution of points shows no particular pattern. Which graph shows which aspect?

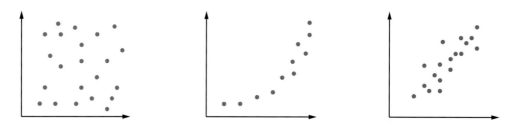

- How many points does our scattergraph involve? The more we have, the more confidence we can have in any model they suggest. (After all, a scattergraph with just two points fits a linear model perfectly but this could not be expected to give a reliable indicator of trend without the evidence of more points suggesting the suitability of a linear model.)

- How closely do the points fit a straight line? The closer the points lie to a straight line, the better the linear model will be and the more confidence we can have in the likely accuracy of our prediction.

- Are we making a prediction for a value that lies within the spread of our known values or is it for a value beyond our known values? I.e., does our prediction involve ***inter*polation** (*between* the given points) or ***extra*polation** (*beyond* the given points). Predictions involving the latter are likely to be less reliable, our confidence in the prediction decreasing the further we are from known points. We often do tend to extrapolate beyond known data points, for example weather forecasts predict the weather in a few days time based on the weather patterns of past days, but the further ahead we are predicting the less certain we are likely to be.

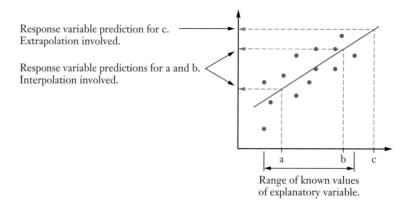

Response variable prediction for c.
Extrapolation involved.

Response variable predictions for a and b.
Interpolation involved.

a   b   c

Range of known values
of explanatory variable.

ISBN 9780170394994

Consider again our previous prediction that for a country with a consumption of a particular foodstuff of 45 kg per person per year, the death rate due to heart disease would be approximately 230 deaths per year per 100 000 people.

We could be reasonably confident in this prediction because:

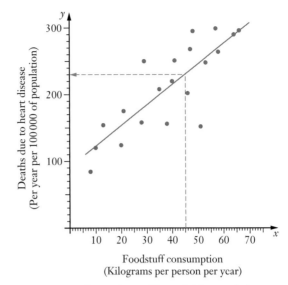

- The graph *does* suggest a linear model could well be appropriate.

- The graph involves a reasonable number of points.

- The figure of 45 kg per person per year is *within* the values covered by the given data, i.e. **interpolation** is involved.

However, we could not be so confident if, for example, we continued our 'line of best fit' beyond the given data to predict the heart disease death rate for a foodstuff consumption figure of 90 kg per person per year for example. In this case **extrapolation** would be involved because we would be going beyond the range of the known data points.

Exercise 1C

1 CAR VALUES

For the 14 cars of a particular make and model advertised in the Motors for Sale section of a newspaper one weekend the asking price and age of the cars were as follows:

Age (years)	13	3	1	7	4	2	11
Price	$6500	$19 300	$20 988	$9900	$16 727	$19 990	$5200

Age (years)	6	2	11	1	6	4	9
Price	$13 500	$16 990	$6250	$21 990	$12 500	$15 500	$9900

Display these figures as a scattergraph with the age on the horizontal axis and the asking price on the vertical axis.

Comment on any relationships shown between the variables and whether a linear (i.e. straight line) model would be suitable.

Draw on your graph the straight line you feel best summarises the data.

Use your straight line to suggest what a likely price was for a 5 year old car of this make and model at the time of the weekend paper.

1. Bivariate data ●●●●●●

2 EXAM RESULTS

Fifteen Mathematics students sat a mid-year exam and an end-of-year exam. The results they gained were as follows:

Initials	LA	TB	JD	TE	FG	LG	IH	GL	KL	AL	BR	JS	RS	PT	LV	Out of
Mid-year	40	34	35	32	36	64	69	63	55	29	59	45	68	66	51	92
End-of-year	50	69	75	68	64	89	116	98	103	46	82	61	121	88	98	135

a Plot a scattergraph of these results with mid-year results on the horizontal axis and end-of-year results on the vertical axis.

b Comment on any trends shown and whether a linear model would be a reasonable one from which to make estimates.

c Include on your graph what you consider to be the 'line of best fit'.

d Another student scored 50 out of 92 in the mid-year exam but was hospitalised at the end of the year due to a road accident and hence missed the end-of-year exam. Use your line of best fit to estimate an end-of-year exam mark for this student.

3 SPHERES

A school science experiment involved students in estimating the volumes of certain spheres. The students used graduated beakers containing water and noted the volume of water displaced when each sphere was immersed in the water. The students also measured the diameter of each sphere.

One student recorded the following results:

Sphere	A	B	C	D	E	F	G	H
Diameter	2 cm	3 cm	4 cm	5 cm	6 cm	7 cm	8 cm	9 cm
Volume	5 cm^3	15 cm^3	35 cm^3	65 cm^3	115 cm^3	180 cm^3	270 cm^3	380 cm^3

a Represent these paired values as points on a graph with diameter on the horizontal axis and volume on the vertical axis.

b Would a linear model be a good choice for this data?

4 BIRTHDATES

Thirty-two year eight students were asked to state their birth dates in the form 'day of month/month'. Thus a response of 5/11 means the 5th of November. The responses for the 32 students are given below.

13/3	13/6	19/5	24/7	24/4	19/3	15/6	11/4
29/1	15/2	10/3	13/12	3/10	18/2	3/1	15/8
1/5	15/3	16/11	25/12	4/7	25/5	29/11	4/12
20/9	16/6	18/6	20/10	30/9	4/3	24/4	24/1

Construct a scattergraph of this data with the day of the month (1, 2, 3, ... , 31) on the horizontal axis and the month (1, 2, 3, ... , 12) on the vertical axis.

Comment on any relationships shown.

5 FUEL CONSUMPTION

Twenty cars were tested for fuel consumption when driving in the city and for highway driving. The figures obtained were as shown below.

	Litres per 100 km		Engine size in litres
	City cycle	Highway cycle	
Vehicle 1	9	6.8	1.8
Vehicle 2	8	6	1.6
Vehicle 3	9.5	6.6	2.0
Vehicle 4	10.5	8	2.5
Vehicle 5	6.4	5.2	1.3
Vehicle 6	12	7.6	4.0
Vehicle 7	14	9	5.0
Vehicle 8	6.8	5	1.3
Vehicle 9	8.5	6.4	1.6
Vehicle 10	8	6.6	1.8
Vehicle 11	10	6.4	2.2
Vehicle 12	8.5	5.6	2.0
Vehicle 13	11	7	3.8
Vehicle 14	14	8.5	5.0
Vehicle 15	14	9	5.0
Vehicle 16	7.2	5.2	1.5
Vehicle 17	10	6.2	2.0
Vehicle 18	10	7.2	3.0
Vehicle 19	10.5	7	2.6
Vehicle 20	10.5	6.6	2.4

a Plot a graph with the highway cycle fuel consumption figures on the horizontal axis and the city cycle fuel consumption figures on the vertical axis. Comment on any relationships indicated by the graph.

b Plot a graph with the engine size on the horizontal axis and the city cycle fuel consumption figures on the vertical axis. Comment on any relationships indicated by the graph.

Imagefolk/Zoonar/P.Gudella

6 COOKING THE MEAT

A cookery book gave the cooking time for a particular joint of meat according to the weight of the joint. The cooking times and weights were as shown in the table below.

Weight	1 kg	1.5 kg	2 kg	2.5 kg	3 kg	3.5 kg	4 kg
Time	1 hour	1 hour 20 mins	1 hour 40 mins	2 hours	2 hours 20 mins	2 hours 40 mins	3 hours

Plot a scattergraph of these figures and comment on any relationships shown.

7 AIRFARES

Let us suppose that the cost of a return flight from Perth to various places in Australia flying business class, together with the distance from Perth to each of these places are as shown in the following table.

Flight	Cost	Distance (one way, by air)
Perth–Adelaide–Perth	$2770	2120 km
Perth–Brisbane–Perth	$3500	3611 km
Perth–Darwin–Perth	$3010	2651 km
Perth–Melbourne–Perth	$3070	2707 km
Perth–Sydney–Perth	$3140	3284 km

a Plot a scatter graph of these figures with distance on the horizontal axis and cost on the vertical axis and comment on any relationships shown.

Assuming suitable flights did exist, use your graph to suggest prices for the following return flights, flying business class, and comment on the likely reliability of your estimates.

b From Perth to a place in Australia that is 2500 km away by air,

c From Perth to a place overseas that is 15 000 km away by air.

ISBN 9780170394994

8 BREAD ON SPECIAL

The manager of a bread shop monitors the number of loaves of a particular size and style they sell each Saturday morning for eight weeks, and the price they charge for a loaf. The data collected was as shown in the table shown below.

Week	1	2	3	4	5	6	7	8
Price	$2.70	$2.35	$3.05	$2.60	$2.50	$2.95	$3.20	$2.80
Loaves sold	342	384	277	331	382	280	254	328

Plot a scattergraph that would allow consideration of how the price per loaf and the number of loaves sold might relate to each other and comment on any relationship shown.

In week 9 the baker plans to charge $2.90 per loaf. How many would you advise him to make if he wants to sell as many as possible but does not want any left?

9 MEMORY

A list of twenty words was read to a group of 28 students who were then asked to recall as many of the words as they could one minute later. The words were read out again and recall was tested again after 15 minutes. The paired scores achieved in these two memory tests are shown below.

Initials	SA	PB	CC	TC	VC	BD	KD	LF	GG	LH	NH	SH	TH	KL
1st score	9	15	11	11	13	10	12	10	11	13	9	7	12	11
2nd score	15	18	14	15	15	17	18	10	11	17	10	9	17	15

Initials	TL	DM	FN	KP	TP	GR	MR	DS	NS	TS	JW	IW	SY	SZ
1st score	10	10	7	7	18	7	9	6	12	9	14	6	14	5
2nd score	15	12	10	6	20	8	15	9	11	11	17	7	18	7

Draw a scattergraph of these results and comment on any trends shown.

Draw what you consider to be 'the line of best fit' and use this line to predict the likely second score for a student with a first score of 8.

Miscellaneous exercise one

This miscellaneous exercise may include questions involving the work of this chapter and the ideas mentioned in the Preliminary work section at the beginning of the book.

1 Solving equations.

   a If $y = 0.4x + 16.7$, find y when $x = 19.5$.   **b** If $y = 4.25x - 17$, find y when $x = 84$.

   c If $y = -7.5x + 307$, find y when $x = 26$.   **d** If $y = 5.2x + 3.7$, find x when $y = 128.5$.

   e If $y = -3.25x + 127$, find x when $y = 10$.

2 The length of a particular spring, y cm, is found to depend upon the weight, w kg, suspended from it, according to the rule $y = 10w + 25$.

Interpret the numbers 10 and 25 in this situation.

3 Copy and complete the following table for the lines A to G shown below.

	Coordinates of y-axis intercept	Gradient	Equation
A			
B			
C			
D			
E			
F			
G			

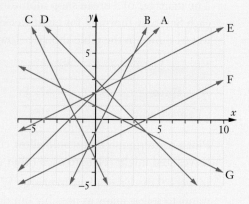

4 A survey into the health and fitness of the Australian population sampled over 9000 men and women of certain ages. The data concerning the heights of these people were as follows:

AVERAGE HEIGHTS OF MALE AUSTRALIANS IN HEALTH SAMPLE

Age (years)	22	27	32	37	42	47	52	57	62	67
Average height (cm)	177	177	176	176	176	175	174	173	173	172

AVERAGE HEIGHTS OF FEMALE AUSTRALIANS IN HEALTH SAMPLE

Age (years)	22	27	32	37	42	47	52	57	62	67
Average height (cm)	164	163	163	163	162	163	161	160	160	159

Display these results graphically and comment on any trends shown.

5 Let us suppose that to investigate whether there might be any association between alcohol consumption and the type of crime a person commits, researchers asked 222 people, who had been convicted of particular crimes, to classify themselves by the amount of alcohol they regularly consumed around the time in their lives that they committed the crime. The classifications of alcohol consumption were:

None,   Light,   Medium,   Heavy.

The results led to the following table:

	Usual alcohol consumption			
	None	Light	Medium	Heavy
Assault	2	6	14	18
Fraud	5	16	10	4
Motor vehicle	11	29	28	17
Theft (not involving assault)	3	22	19	18

Investigate whether the proportions of each type of crime change much as we move across the categories of alcohol consumption. Do the figures suggest there is an association between alcohol consumption and the type of crime committed? If such an association appears to exist write some comments describing its nature.

Bivariate data –
further analysis

In the previous chapter, after the work involving *categorical data*, we investigated whether *relationships, associations* or *correlation* existed between two *numerical* variables. The method was to plot a scattergraph of paired values to see if the scatter of points suggested some linear, or other, model could satisfactorily summarise what was going on. I.e., we considered what **form** (linear/non linear) any association might take.

Let us now try to be more *descriptive* and *quantitative* about situations for which a linear model seems appropriate.

Suppose that ten students took both a physics test and a mathematics test and the scores they obtained were as shown in the following table:

Physics score	6	9	11	12	14	14	16	16	17	19
Maths score	19	12	27	36	42	34	32	27	44	41

The scattergraph for these scores is shown below left. The fact that the points tend to form an oval shape from lower left to upper right, as shown below right, indicates that for these students, those who did well in the physics test tended to do well in the mathematics test, and those who scored the lower marks in the physics test tended to score the lower marks in the mathematics test

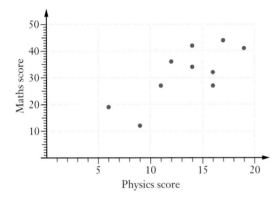 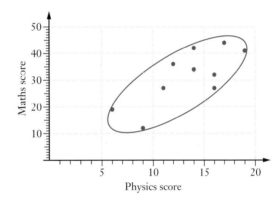

We say that the scattergraph suggests that a **positive correlation** exists between the physics scores and the maths scores.

If a positive correlation exists between two variables, then as one variable increases the other variable increases. Our 'line of best fit' will have a positive gradient.

If instead the points had formed an oval shape from upper left to lower right, that would have suggested a **negative correlation** existed between the variables.

If a negative correlation exists between two variables, then as one variable increases the other variable decreases. Our 'line of best fit' will have a negative gradient.

As well as talking about the **form** (linear or non-linear) of the relationship between two variables and the **direction** of the relationship (positive or negative) we can also talk about its **strength** (strong, moderate or weak).

Positive and negative linear relationships of various strengths are shown below.

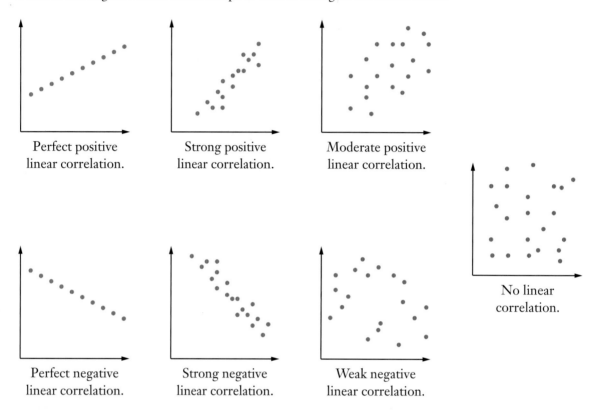

We would first enter these values into a calculator – a typical display is shown on the next page:

Can we do better than 'eye-balling' the line of best fit?

In chapter one we drew lines of best fit by 'eye-balling' the scattergraph and, if a linear model seemed appropriate, drawing in the straight line that seemed to best fit the data.

The weakness in this process is drawing the line of best fit 'by eye'. What one person considers the best line may be quite different to the line chosen by someone else. Indeed it is easy to find yourself questioning your own choice of line moments after drawing it!

Thankfully many calculators can determine the line considered to be *the* best straight line, draw it, give its equation and use it to predict values.

Returning to the physics and maths test results of the previous page:

Physics score	6	9	11	12	14	14	16	16	17	19
Maths score	19	12	27	36	42	34	32	27	44	41

We would first enter these values into a calculator – a typical display is shown on the next page:

L1	L2	L3
6	19	
9	12	
11	27	
12	36	
14	42	
14	34	
16	32	

L2(1)=19

The calculator can then use **linear regression** to:

- Output the equation of the line of best fit in the form $y = mx + c$ where y is the maths score and x is the physics score (see display below left).

- Display the points and the line of best fit (see display below right).

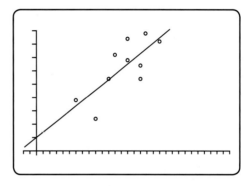

LinearReg
y = ax + b
a = 1.982905983
b = 4.829059829
r^2 = 0.5748032337
r = 0.7581577894

Line of best fit
$y = 1.98x + 4.83$

- Use this line of best fit to predict a score in the maths test based on a score in the physics test. As shown on the right, for a physics score of 13.

  Remember though that if **extrapolation** is involved (rather than **interpolation** as used here) any prediction would have to be viewed with caution.

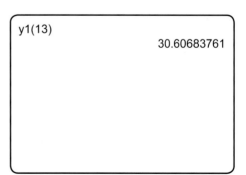

y1(13)

30.60683761

Note: The calculator display above left also shows values it calls r and r^2. We will give these quantities further consideration later in this chapter. For now though let us concentrate our attention on the line of best fit and using it to make predictions.

Make sure that *you* can use *your* calculator to determine the line of best fit and the predicted value for the above bivariate data.

Can you get your calculator to display the scattergraph complete with the line of best fit?

Note: To make a distinction between the observed y values in our original data and the predicted y values, obtained from our regression line, the latter are sometimes written as $\hat{y}$, pronounced 'y hat'.

Using this notation for the predicted value of y on the previous page:

$$\text{when} \quad x = 13, \quad \hat{y} = 30.6$$

Similarly this 'hat notation' can be used when stating the equation of the regression line:

$$\hat{y} = 1.98x + 4.83$$

This emphasises that the equation gives a predicted y value ($\hat{y}$) from a known x value.

Spreadsheets

Spreadsheets can also be used to display and analyse bivariate data in this way:

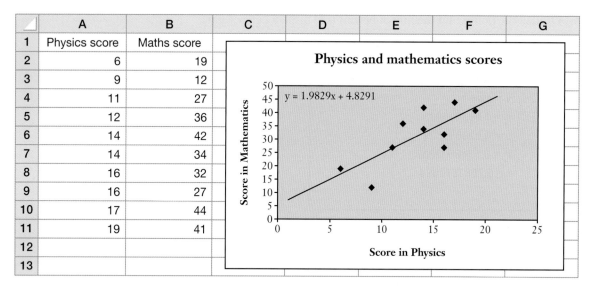

Try to use a computer spreadsheet yourself to display such data graphically.

Exercise 2A

1 An investigation was carried out to see how the time (t seconds) a child took to complete a particular task depended on the age (y years) of the child ($5 \leq y \leq 18$).

The data collected suggested that the situation could be well modelled using a line of best fit with equation:

$$t = -1.5y + 41$$

Interpret the numbers −1.5 and 41 in this situation.

2 A study investigated the relationship between the time, h hours, that a student revises for an exam, and the mark, $m\%$, gained in the exam. The results suggested that the situation could be reasonably well modelled using the rule:

$$m = 2h + 28.$$

Interpret the numbers 2 and 28 in this situation.

3 The equations of the lines of best fit for the five scattergraphs A to E, shown below, are in the equations list also shown below.

For each scattergraph match the appropriate equation of the line of best fit. (You should be able to do this 'by inspection', not by obtaining the line of best fit equation from your calculator.)

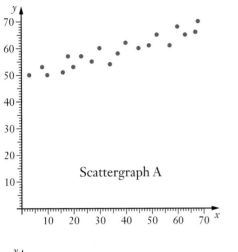

Scattergraph A

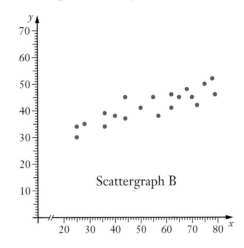

Scattergraph B

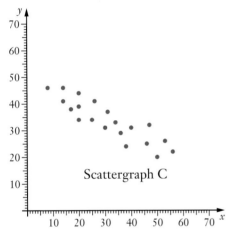

Scattergraph C

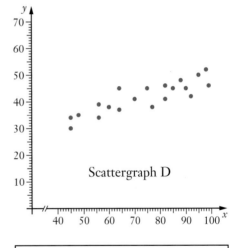

Scattergraph D

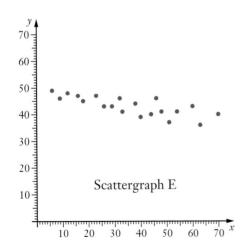

Scattergraph E

Equations

$y = -0.490x + 49.0$

$y = -0.152x + 48.6$

$y = 0.271x + 48.9$

$y = 0.283x + 26.4$

$y = 0.283x + 20.7$

4 Seventeen students following a particular course had to take two tests, the first out of 40 and the second out of 25. One of the students was in hospital when the second test occurred and could not sit the test.

The marks obtained by the students are given below.

1st test	4	11	12	12	16	20	21	21	27	26
2nd test	11	9	15	4	16	13	16	20	18	Abs

1st test	28	30	30	35	35	39	39
2nd test	22	17	20	22	25	19	20

Draw a scatter diagram showing the results of the sixteen students who completed both tests.

Draw the line of best fit 'by eye' and hence estimate a second test mark for the student who was unable to sit that test.

Use a spreadsheet or your calculator to create a scattergraph of the data complete with the line of best fit and compare it to your scattergraph and your line of best fit.

Determine the equation of the line of best fit from your calculator or spreadsheet.

Determine an estimated second test score for the student who was unable to sit that test using the calculator or spreadsheet's line of best fit and compare it to your 'by eye' prediction.

5 Let us suppose that a medical check-up involves, amongst other things, measuring something called the *peak expiratory flow rate*, to detect possible respiratory problems. The peak flow rates achieved by twenty males, all aged between 30 and 50, are given in the table below, together with the height of each of these males.

Height (cm), h	184	189	171	166	200	192	176
Peak flow rate (units/min), P	500	555	510	550	625	610	525

Height (cm), h	162	188	164	180	168	185	176
Peak flow rate (units/min), P	500	570	475	530	520	575	570

Height (cm), h	189	194	195	190	191	172
Peak flow rate (units/min), P	525	600	560	590	585	540

Draw a scattergraph of these figures with the height, h, on the horizontal axis and peak flow, P, on the vertical axis and draw the line of best fit 'by eye'.

Hence estimate the peak flow rate for a male aged between 30 and 50 and of height 174 cm.

Now use a calculator or spreadsheet to display the data as a scattergraph, obtain the equation of the line of best fit and the predicted peak flow rate for a male aged between 30 and 50 and of height 174 cm and compare the value with your 'by eye value'.

ISBN 9780170394994

6 Sixteen students following a particular course had to take two examinations, one in mathematics and one in physics. One of the students was taken ill during the physics exam and could not complete the exam.

The marks obtained by the students are given below.

Initials	JA	FC	ED	SF	JG	KK	ML	TR
Maths (%)	13	87	55	30	63	11	68	40
Physics (%)	30	49	35	41	ill	20	60	34

Initials	RR	RS	JS	WT	AV	ZW	LW	PY
Maths (%)	85	30	58	72	40	20	95	45
Physics (%)	56	25	43	52	55	27	67	54

Either by drawing it, or by using a calculator or spreadsheet, view a scattergraph showing the results of the fifteen students who completed both tests with M, the maths percentage on the horizontal axis, and P, the physics percentage, on the vertical axis.

Comment on the suitability of using a linear model.

Determine the equation of the line of best fit in the form $P = aM + b$ and use this to determine an estimated physics mark for the student who was unable to complete that exam.

7 The table below gives the *Infant Mortality Rate* (deaths per 1000 infants) and the *Literacy Rate* (percentage of population who are literate) for 15 countries.

Country	A	B	C	D	E	F	G	H
Literacy rate (%), L	64	81	52	87	78	95	35	99
Infant mortality, I	43	60	78	27	52	29	107	9

Country	I	J	K	L	M	N	O
Literacy rate (%), L	93	77	52	98	35	69	42
Infant mortality, I	15	67	76	17	102	74	90

a With the literacy rate on the horizontal axis, and the mortality rate on the vertical axis, view the data as a scattergraph either by drawing it or by displaying it on a calculator or spreadsheet.

b Determine the equation of the line of best fit in the form $I = aL + b$.

c Hence determine an estimate of the infant mortality rate for a country

  i with a literacy rate of 48%,

  ii with a literacy rate or 15%,

and in each case comment on how reliable your predictions might be.

Pearson's correlation coefficient

Describing a relationship between two variables using terms such as strong positive or weak negative is rather imprecise. In this section we will see how the strength of a relationship can be stated in terms of a **correlation coefficient**, a number from −1 to 1 (also known as Pearson's correlation coefficient).

In the calculator display shown on an earlier page, and shown again on the right, 'r' is the correlation coefficient.

The display also gives r^2, the square of the correlation coefficient, called the *coefficient of determination*. We will consider this later but for now let us concentrate our attention on the correlation coefficient, r.

LinearReg
y = ax + b
a = 1.982905983
b = 4.829059829
r^2 = 0.5748032337
r = 0.7581577894

The **correlation coefficient** is a measure of how closely the relationship between two sets of data approximates to a linear relationship. We tend to use r for the correlation coefficient where $-1 \leq r \leq +1$. We sometimes write the correlation coefficient as r_{xy} where the subscripted letters are the two variables involved, in this case x and y.

A correlation coefficient of −1 indicates a perfect linear relationship with a negative gradient (as one variable increases the other decreases) and a correlation coefficient of +1 indicates a perfect linear relationship with a positive gradient (as one variable increases the other increases).

A few scattergraphs and the respective correlation coefficients are shown below.

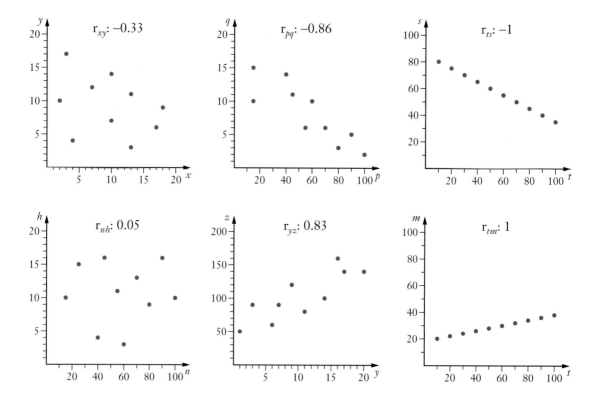

Note:
- Questions in this text may deliberately involve only about ten data points or so to allow you to practise the techniques involved without spending too much time processing the data. In reality, to make more valid any comments as to whether two variables show correlation, we would try to involve more data points.

- The graphs on the right both show points that lie in a perfect straight line but in both cases the idea of a correlation coefficient is meaningless because we do not have *bi*variate data. In each case either x or y is not showing any variation at all.

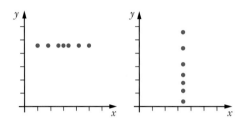

Using your calculator to determine r

As we have already seen, the value of r and the equation of the line of best fit can be determined using a calculator, or computer, that can perform **linear regression**, the name given to the process of determining the best straight line for bivariate data. (The line of best fit is sometimes referred to as the **regression line**.)

For the following bivariate data the value of r and the equation of the linear regression line are as given below the table.

x	92	23	68	64	57	81	12	98	32	20	47	96	81	40
y	51	32	39	34	25	36	21	36	21	20	35	43	42	26

Correlation coefficient, r_{xy} 0.83 (correct to 2 decimal places)
Line of best fit is $y = ax + b$ where $a = 0.262$ and $b = 17.76$.

For $x = 15$ the line of regression gives a predicted y value of 21.7.
For $x = 55$ the line of regression gives a predicted y value of 32.2.
For $x = 88$ the line of regression gives a predicted y value of 40.8.

The graph below shows the bivariate data, the regression line and the predicted values.

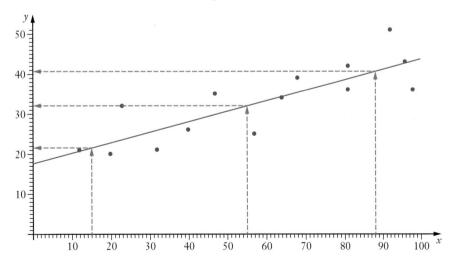

Predictions and residuals

Can we have confidence in our predictions?

Notice that all three of the predicted values on the previous page involve interpolation – a term introduced in the previous chapter.

Hence with

- the distribution of the data points suggesting that a linear model would be appropriate,
- the value for r of 0.83 confirming the presence of a reasonably strong positive linear correlation,
- a reasonable number of points being involved,

and • the fact that interpolation is involved (between known data points), rather than extrapolation (beyond the known data points)

we can be reasonably confident of the validity of our predicted values.

EXAMPLE 1

Ten people were given two tests, A and B and the marks obtained were as follows:

Test A	1	2	2	3	5	5	8	10	12	12
Test B	14	17	18	19	23	25	29	30	35	40

a Plot the data as a scattergraph with the test A results on the horizontal axis and the test B results on the vertical axis.

b Suggest whether or not the data seems suited to the use of a linear model.

c Determine the correlation coefficient for the 10 pairs of results.

d Determine the equation of the line of best fit in the form

$$\text{Test B score} = a \times \text{Test A score} + b$$

e Predict the Test B score for someone with Test A score of 7 and comment on the validity of your prediction.

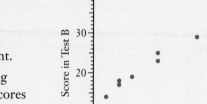

Solution

a The required scattergraph is shown on the right.

b The points on the scattergraph suggest a strong positive linear correlation exists between the scores on test A and the scores on test B. Hence the data seems well suited to the use of a linear model.

c Putting the given data pairs into a calculator we obtain r = 0.978 (3 decimal places).

   (Further confirming the suitability of a linear model.)

d The line of best fit has equation: Test B score $= 1.956 \times$ Test A score $+ 13.26$

e For a test A score of 7, predicted test B score $= 1.956 \times 7 + 13.26$
   ≈ 27

With a reasonable number of points involved, the high value of r and interpolation involved we can be reasonably confident that our predicted value is valid.

ISBN 9780170394994

Cause and effect

Though a scattergraph, and an r value, may suggest that a correlation exists between two variables we cannot assume that the changes in one variable *cause* the changes in the other variable.

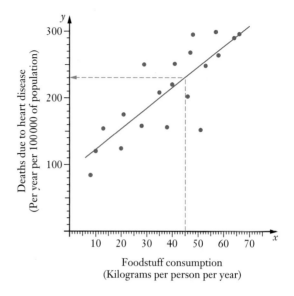

Foodstuff consumption
(Kilograms per person per year)

In an earlier example we found there to be a correlation, or association, between the consumption of a particular food stuff and the death rate due to heart disease. However this does not prove that the foodstuff consumption *causes* the increased death rate. The graph suggested a relationship, not a 'cause and effect'. There could be other factors at work. For example it could be that the high intake of the foodstuff indicates a more affluent society and it is the lifestyle of the more affluent society that is the reason for the higher death rate due to heart disease.

This *affluence of society* is a **lurking variable** or **confounding variable** that each of our two variables, food consumption and heart disease death rates, could have a **common response** to.

To take another example, we would probably not be surprised if the monthly sales of ice-cream and the monthly number of deaths due to drowning were correlated. However in this case the lurking variable, or confounding variable, is likely to be *temperature*. On hot days more ice-creams are sold and more people go swimming, hence the correlation between ice-cream sales and deaths due to drowning. It would be incorrect to conclude that because there is correlation, one is causing the other.

It is also possible that collected data could suggest an association between two variables but the apparent association is purely **coincidental**. For example a person might find that on several occasions, when they buy some seedlings for planting in the garden, the weather for the next few days after planting seems to be very hot, causing the seedlings to die. However this is likely to be just coincidence and certainly not some causal relationship between the seedlings being planted and a spell of hot weather following. A company making cars might find that sales of their cars have been dropping ever since a particular employee started working for the company, but again this is likely to be coincidence, not any causal relationship between the sales figures and the length of time the employee has been with the company.

> **Hence we need to recognise that an observed association or correlation between two variables does not necessarily mean that there is a causal relationship between the variables.** There might be a causal relationship but we cannot assume it exists purely on the basis of there being an association.

ISBN 9780170394994 **2.** Bivariate data – further analysis ●●○○○○

Investigating
relationships between
two numerical
variables

Strong, moderate or weak relationship?

For what values of r do we describe the linear relationship between two variables as being strong, moderate or weak?

It is not quite as clear cut as setting rules like 'for r between 0.3 and 0.5 we say that the linear correlation is weak positive'. Other aspects may need consideration.

How many data points were used to determine r?
Are 'outliers' lowering the value of r?
Some environments, for example medical, may require a higher value of r
before wanting to suggest a 'strong' relationship exists.

To attempt to give some idea of when terminology such as strong, moderate or weak might be considered reasonable, for various r values, the diagram below gives a 'rule of thumb' that readers might consider useful. However, as an internet search would reveal, there are many different suggestions for where the boundaries should lie.

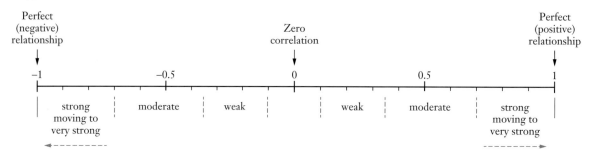

Use the internet to look up what the phrase 'rule of thumb' means.
What is thought to be the origin of the phrase?

About the correlation coefficient, r$_{xy}$

Whilst you should understand what the correlation coefficient tells you about bivariate data you would not normally be expected to calculate its value yourself, preferring instead to obtain its value using the statistical capability of a calculator. However you should be able to read and understand the next few paragraphs to get some understanding of how the correlation coefficient is calculated.

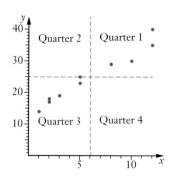

The scattergraph shown on the right shows the bivariate data of example 1, with test A scores shown on an x-axis and test B scores on a y-axis.

Dashed lines are drawn at the mean of the x scores, $\bar{x}$, and the mean of the y scores, $\bar{y}$.

These dashed lines divide the graph into 4 quarters.

For a strong positive linear relationship we would expect most points to lie in quarters 1 and 3, which we can see is indeed the case for this scattergraph.

For points in these quarters the quantities (x score $- \bar{x}$) and (y score $- \bar{y}$) are either both positive or both negative.

Thus the mean of all the products (x score $- \bar{x}$)(y score $- \bar{y}$) will be positive.

For bivariate data with a strong negative linear relationship many of the points will lie in quarters 2 and 4 where the quantities (x score $- \bar{x}$) and (y score $- \bar{y}$) are of opposite sign. The mean of all the products (x score $- \bar{x}$)(y score $- \bar{y}$) will then be negative.

Thus this process gives a number that is positive for bivariate data showing a positive linear relationship and negative for data showing a negative linear relationship.

We call this mean of the products (x score $- \bar{x}$)(y score $- \bar{y}$) the **covariance** of x and y, written s_{xy}.

Hence if we have n data points (x_1, y_1), (x_2, y_2), (x_3, y_3), ..., (x_n, y_n).

$$\text{Covariance, } s_{xy} = \frac{(x_1 - \bar{x})(y_1 - \bar{y}) + (x_2 - \bar{x})(y_2 - \bar{y}) + \ldots + (x_n - \bar{x})(y_n - \bar{y})}{n}$$

For the ten pairs of scores on the scattergraph on the previous page $\bar{x} = 6$ and $\bar{y} = 25$.

For these points:

x	1	2	2	3	5	5	8	10	12	12	
y	14	17	18	19	23	25	29	30	35	40	
$x - \bar{x}$	−5	−4	−4	−3	−1	−1	2	4	6	6	
$y - \bar{y}$	−11	−8	−7	−6	−2	0	4	5	10	15	Total
$(x - \bar{x})(y - \bar{y})$	55	32	28	18	2	0	8	20	60	90	313

$$\text{Covariance, } s_{xy} = \frac{313}{10}$$
$$= 31.3$$

However the magnitude of this covariance will depend on the magnitude of the original x and y values. To overcome this we then divide by the standard deviations of the x and y sets of data. (The size of these quantities are similarly dependent upon the size of the original data and so division will cancel this effect.) This will give a number that is positive for bivariate data showing a positive linear relationship, negative for data showing a negative linear relationship, and will be independent of the size of the original data. This number is the **correlation coefficient**, r_{xy}, and is such that $-1 \le r \le 1$.

$$r_{xy} = \frac{\text{Mean of the products: } (x \text{ score} - \bar{x})(y \text{ score} - \bar{y})}{(\text{Standard deviation of the } x \text{ scores})(\text{Standard deviation of the } y \text{ scores})}$$

$$\text{i.e. } r_{xy} = \frac{s_{xy}}{s_x s_y}$$

$$= \frac{31.3}{(4)(8)} \text{ (for the data given above)}$$

$$= 0.978125$$

Confirm that this is the value your calculator gives for the correlation coefficient for this set of ten data pairs.

About the line of best fit

The scattergraph on the right involves ten points and as well as showing the points, and their line of best fit, it also shows the vertical deviations of each point from the line of best fit.

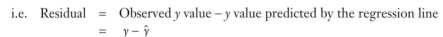

These lines show the differences between the obtained y values and the y values predicted by the regression line.

These differences are called the linear regression **residuals**.

i.e. Residual $=$ Observed y value $- y$ value predicted by the regression line
$ = y - \hat{y}$

When a calculator determines the line of best fit for estimating y for given values of x it is finding the line which minimises the sum of the squares of these vertical deviations.

We call this the line of best fit, the trend line, or the

$$\textbf{least squares regression line of } \boldsymbol{y} \textbf{ on } \boldsymbol{x}.$$

It is used to predict $\boldsymbol{y}$ **based on** $\boldsymbol{x}$.

Note
- The fact that our regression line is determined by minimising the sum of the *squares* of the deviations of each point from the line means that any outlying points can greatly influence the location of the regression line.

- Prior to the availability of calculators capable of processing statistical data the least squares regression line was determined using the rule:

  Regression line, y on x, is $\qquad y - \overline{y} = \dfrac{s_{xy}}{s_x^2}(x - \overline{x}).$

- Notice that if we substitute $x = \overline{x}$ into the regression line equation from the previous dot point:

  $$y - \overline{y} = \frac{s_{xy}}{s_x^2}(\overline{x} - \overline{x})$$

  i.e. $\qquad\qquad y - \overline{y} = 0$

  and so $\qquad\qquad\quad y = \overline{y}$

  Thus:

The least squares regression line passes through $(\overline{x}, \overline{y})$.

- We predict y using the y **on** x **regression line**, a line determined by minimising the sum of the squares of the *vertical* deviations of each point from the line. We should *not* use the equation of this y on x line 'in reverse' to estimate an x value given a y value. Instead we use the x **on** y **regression line**, a line determined by minimising the sum of the squares of the *horizontal* deviations.

> Hence the y **on** x **regression line** is used to predict y **based on** x and the x **on** y **regression line** is used to predict x **based on** y.

With the explanatory variable as our x values and the response variable as our y values we would usually want to predict y, given x. Hence it will usually be the y on x regression line that we will use.

For r close to ±1 there will be little difference between these lines and they will coincide if r = ±1.

The graph shown below shows both regression lines for the following data, encountered earlier in this chapter:

x	92	23	68	64	57	81	12	98	32	20	47	96	81	40
y	51	32	39	34	25	36	21	36	21	20	35	43	42	26

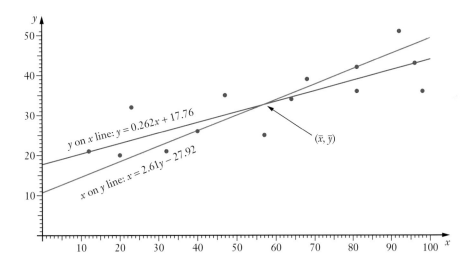

Make sure you can obtain the equation of the x on y regression line.

(Instruct your calculator to read the ordered pairs as (y, x) when it calculates the line of best fit).

Exercise 2B

1 Suppose that a survey of teenagers found that there was an association between the number of hours spent reading teen fashion magazines and the chance of the teenager having an eating disorder.

Can we conclude that the reading of fashion magazines is causing the eating disorder?

Write a few sentences explaining what this association might be due to.

2 From the time that a particular individual started his employment as head gardener at the XYZ export factory the company profits have been in steady decline. As the length of time that the individual has been with the company has increased then so the profits have decreased.

Can we conclude from this association between length of employment of the individual and company profits that it is the individual who is causing the decline in company profits?

Write a few sentences discussing this situation.

3 A survey found that over recent years the average age of school leavers had shown a steady increase and so too had the number of school leavers having attention deficit disorder.

Could we conclude from this association between years at school and cases of attention deficit disorder that it is the extra time at school that is causing the attention deficit disorder?

Write a few sentences discussing this situation.

For the data given in each of questions 4 to 11 use your calculator to determine:

a the correlation coefficient r_{xy},

b the equation of the regression line of y on x: $y = Ax + B$,

c the predicted y values requested in each question and state for each predicted value whether it is *interpolation* or *extrapolation* that is involved.

(For questions 4 to 7 assume integer coordinates.)

4

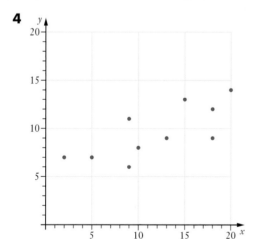

Predict the y value for $x = 28$.

5

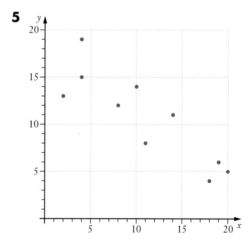

Predict the y value for $x = 6$.

ISBN 9780170394994

6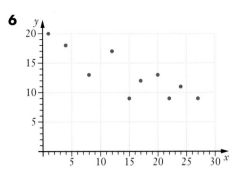

Predict the *y* value for **i** *x* = 10,

                            ii *x* = 40.

7

Predict the *y* value for **i** *x* = 40,

                            ii *x* = 17.

8

x scores	*y* scores
1	2.7
2	3.1
3	2.5
4	2.5
5	3.1
6	3.4
7	3.2
8	3.7
9	3.1
10	3.5

Predict the *y* value for *x* = 11.

9

x scores	*y* scores
18	43
40	30
25	28
4	50
54	15
45	15
59	5
33	14
54	10
10	30

Predict the *y* value for **i** *x* = 22,

                            ii *x* = 47.

10

x scores	*y* scores
4.5	340
1.7	320
4.1	360
5.7	500
2.5	280
7.2	520
3.7	410
6.5	530
3.2	410
1.3	280
0.8	300
5.6	390

Predict the *y* value for **i** *x* = 5.1,

                            ii *x* = 9.2,

                           iii *x* = 12.9.

11

x scores	*y* scores
95	65
50	29
20	5
120	85
35	17
65	41
110	77
70	45
45	25
30	13
60	37
105	73

Predict the *y* value for **i** *x* = 85,

                            ii *x* = 40,

                           iii *x* = 200.

For the data given in each of questions 12 and 13 below determine

a the equation of the regression line of q on p: $q = ap + b$.

b the equation of the regression line of p on q: $p = cq + d$.

c the predicted values requested using the appropriate regression line.

(For question 12 assume integer coordinates.)

12

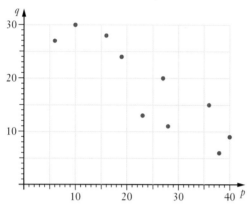

Predict **i** the q value for $p = 30$,

ii the p value for $q = 25$.

13

p scores	q scores
35	19
20	16
60	24
45	21
5	13
95	31
25	17
85	29
55	23

Predict **i** the q value for $p = 48$,

ii the p value for $q = 32$.

14 The owner of a fish and chip shop feels sure that on any summer's evening the number of customers buying fish and chips is dependent upon the temperature that evening. To test this theory one summer he notes the temperature at 5 pm when he opens his shop and also notes the number of serves of fish he sells that evening.

A period of 15 summer evenings produced the following results:

Evening	1	2	3	4	5	6	7	8
Temp (T°C)	31	35	24	33	34	25	37	33
No. of serves (N)	202	83	208	141	128	157	72	112

Evening	9	10	11	12	13	14	15
Temp (T°C)	30	27	31	29	34	33	28
No. of serves (N)	142	203	104	170	99	157	121

a Determine the correlation coefficient r_{TN}.

b Determine the equation of the line of regression $N = A + BT$.

c Predict the value of N for

i $T = 26$

and **ii** $T = 42$

commenting on the likely validity of your predictions in each case.

15 The shop keeper of question 14 feels that during winter the trend is different. The lower temperatures now make going out to get dinner rather unpleasant and people are less likely to visit his shop on very cold evenings, preferring instead to use a food outlet that has a delivery service (which his does not). Again he notes the temperature at 5 pm when he opens his shop and the number of serves of fish he sells that evening. A period of 15 winter evenings produced the following results:

Evening	1	2	3	4	5	6	7	8
Temp (T°C)	2	16	15	8	6	19	4	9
No. of serves (N)	40	153	192	104	98	180	52	130

Evening	9	10	11	12	13	14	15
Temp (T°C)	10	3	13	17	7	11	6
No. of serves (N)	173	91	230	242	51	130	131

a Determine the correlation coefficient r_{TN}.

b Determine the equation of the line of regression $N = A + BT$.

c Predict the value of N for

      i $T = 14$

and   **ii** $T = -3$

commenting on the likely validity of your predictions in each case.

16 In a particular course sixteen students sat a mid-year exam, marked out of 80, and an end of year exam, marked out of 120. The results they obtained were as follows:

Initials	TB	KD	MF	KH	TH	LK	PL	RL
Mid year (out of 80)	47	72	54	32	78	53	65	74
End year (out of 120)	82	98	92	56	105	81	79	115

Initials	SM	TN	WN	KS	RS	TT	VW	PY
Mid year (out of 80)	52	53	42	67	60	67	43	66
End year (out of 120)	71	84	73	104	104	88	54	100

a Determine the correlation coefficient for the above set of results.

b Determine the equation of the least squares regression line for predicting end of year marks based on mid year marks.

c Another student scored 63 out of 80 in the mid year exam but did not sit the end of year exam due to a family bereavement. Use your regression line to estimate an end of year exam mark for this student.

17 Scientists are testing a new spray for its effectiveness in controlling a particular weed that is toxic to sheep. The scientists test various concentrations of the spray over test areas of 1 hectare in which the weed is prevalent. The scientists then determine the percentage of the weed in this 1 hectare test area that survives the spray one week after its application. The results obtained are given in the table below.

Concentration (C%)	10	25	30	45	50	55	60	70	75	80
Survival rate (S%)	32	23	24	20	18	12	15	9	8	2

a Determine the correlation coefficient r_{CS} and the equation of the least squares regression line for determining S based on C.

b It is likely that if the spray is allowed into widespread use the maximum concentration allowed will be 38%, for health reasons.

If spray of this level of concentration is applied to an area in which the weed is prevalent, estimate the percentage of the weed which would not survive the spray one week after its application.

Shutterstock.com/Masson

18 The table below gives the life expectancy in years for ten countries and also the percentage of the population of those countries having access to good sewage disposal and toilet facilities (i.e. good sanitation facilities).

Country	A	B	C	D	E	F	G	H	I	J
Percentage with access to good sanitation (S%)	26	29	31	43	53	75	87	92	94	97
Life expectancy (L yrs)	59	61	52	62	63	71	67	68	76	70

a Determine the correlation coefficient r_{SL} and the equation of the least squares regression line for determining L based on S.

b Hence predict the life expectancy for a country with

i 60% of its population with access to good sanitation facilities,

ii 80% of its population with access to good sanitation facilities.

c Can we conclude that a high life expectancy is due to having access to good sanitation?

ISBN 9780170394994

Some further considerations

1 Outliers

Consider the scattergraph shown on the right.

The point (40, 10) seems to be very much out of step with the reasonably strong linear trend suggested by the other points.

The point (40, 10) is an **outlier**.

Outliers can significantly alter the location of the line of best fit. The two versions of the scattergraph shown below include the line of best fit with the outlier included in the determination of the line, below left, and with the outlier excluded, below right.

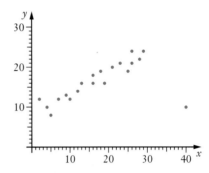

Notice how the predicted y-value for an x-value of 27 differs in the two graphs.

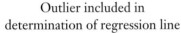

Outlier included in determination of regression line

Outlier not included in determination of regression line

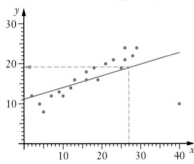

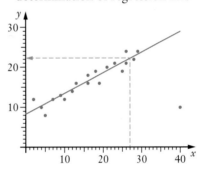

Determining the line of best fit involves finding the straight line that best fits the data points and summarises the overall trend of the points. The presence of an outlier can greatly affect the calculation of the equation of this line. An outlier can have a greater effect on the determination of the line of best fit than a single point should have. To get a straight line that best summarises most of the points we may choose to leave the outlier out when determining the line of best fit. However it is important to realise that whilst in this section we are suggesting that an outlier should be removed to avoid it having an undue influence on the determination of the line of best fit, we are not suggesting that whenever there are outliers in statistical data, remove them. In some areas of investigation the one response that seems out of line with all of the others could be the very thing we are searching for. Consider for example a research worker testing many different drugs to see which, if any, have any effect on a certain disease. In this case the one drug that produces results that are different to all of the others could be the very one that holds the key to defeating the disease. The response should be to *investigate outliers further* rather than just reject them without thought. However with scattergraphs, if we are wanting to determine a line that best summarises the overall trend suggested by the majority of the points, we may need to remove outliers from the calculation to avoid them having too significant an effect on the line of best fit.

Do a search on the internet to investigate how the automatic rejection of outliers apparently delayed the detection of the thinning of the ozone layer.

ISBN 9780170394994

2 'Trimming' or 'cropping' bivariate data

Given a set of data containing an outlier we may choose to remove the outlier when analysing the data. This could be viewed as *trimming* or *cropping* the data and 'harvesting' only some of it for analysis.

Consider the scattergraph on the right.

The strong linear trend suggested for the data points with the smaller x values seems not to apply at all for x values over 25.

If we wanted to make predictions based on x values in the range 3 to 25 we could crop the data so that it only included points for which

$$3 \le x \le 25$$

and determine and use the line of best fit for this cropped data only.

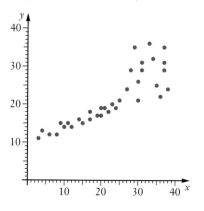

Regression line determined using all data points

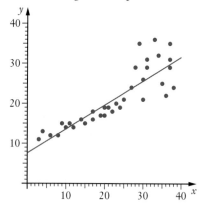

Regression line determined using only data points for which $3 \le x \le 25$

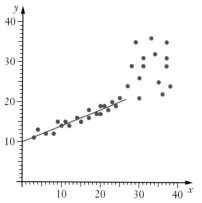

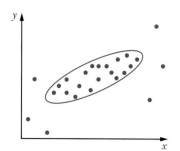

The scattergraph shown on the left seems to have a number of outliers for the lower and higher values of x but the data points in the middle range of x values seem to have quite a strong linear correlation.

In this case we might consider cropping the data so that our line of best fit is calculated using only those data points lying in the oval shown on the graph.

Similarly, for the scattergraph shown on the right, we might consider rejecting the four outlying points and only consider the trimmed or cropped data lying in the oval shown.

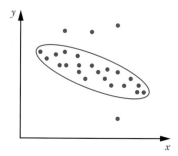

ISBN 9780170394994

3 Residual plots

So far this chapter has concentrated on determining the *straight* line that best fits bivariate data. We have used *linear* regression techniques to model sets of ordered pairs (x, y) in an attempt to predict the likely value of one variable given the value of the other. However, will a straight line always be the best choice of regression model?

Consider the three scattergraphs given below.

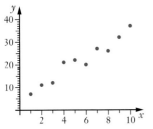

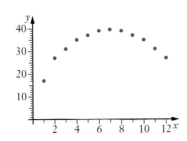

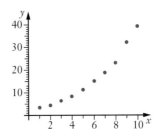

Shape suggests that a linear model would be appropriate.

Shape suggests that a linear model would not be appropriate. Perhaps a quadratic model would be more appropriate.

Shape suggests that a linear model would not be appropriate. Perhaps an exponential model would be more appropriate.

Notice also that if we were to attempt to fit a straight line to each of the above sets of bivariate data the **residuals** (observed y value minus $\hat{y}$, the value predicted by the regression line) are reasonably random for the data set for which linear regression is appropriate but form a pattern in the other two.

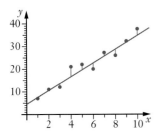

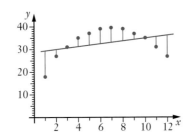

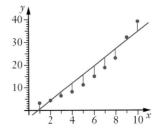

Residuals seem to be random.

Residuals are increasingly negative for high and low values of x and have positive values for the 'middle' x values.

Residuals are negative for the 'middle' values of x and positive for high and low values.

This may be more apparent by considering the **residual plots** shown below.

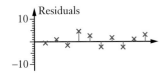

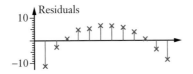

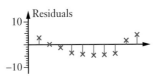

If we did not have the initial graphs to look at but did have a table of residuals, or a residual plot, then noticing the random nature of the residuals can be an indication that a linear model would be appropriate (or a non-random set of residuals indicates that a linear model may not be the most appropriate).

Interpreting regression lines and the coefficient of determination

4 r^2, the coefficient of determination

A question encountered earlier in this book involved the advised cooking time for a joint of meat, according to the weight of the joint. The scattergraph for the situation is shown on the right. Notice that there is a perfect linear association between the cooking time and the weight of the joint. The correlation coefficient, r, is 1.

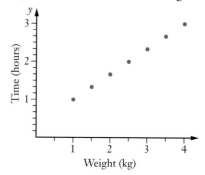

Scattergraph of cooking times for joints of meat of various weights

The variation in the cooking time is 100% explained, or accounted for, by the relationship between the two variables, cooking time and weight of joint. Knowing the weight of the joint we can exactly determine the advised cooking time.

Now let us again consider the food consumption / heart disease death rate situation encountered previously.

For this data

Least squares regression line:

$$y = 3.1x + 90.9$$

Correlation coefficient, r (to 3 decimal places):

0.823

Coefficient of determination, r^2 (to 3 decimal places):

0.677

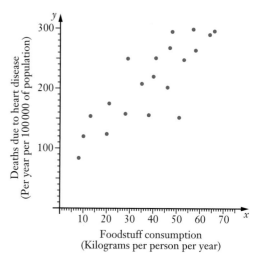

The variation in death rate due to heart disease is not 100% explained by the variation in the foodstuff consumption. However we can use r^2 to state what proportion of the total variation in the death rate due to heart disease is explained, or accounted for, by the variation in the foodstuff consumption. With an r^2 value of 0.677 we can say that approximately 68% of the variation in death rates due to heart disease can be explained, or accounted for, by the variation in the foodstuff consumption.

Note carefully that we are saying 'explained by', not 'caused by'. There is an association between the two variables, so as one changes we should expect the other to change, but that is not to say that the changes in one variable cause the changes in the other.

> If a linear association exists between two variables such that the correlation coefficient is r, then r^2, the coefficient of determination, is the proportion of the variation that can be explained by the linear relationship.

For example suppose that a survey found that the correlation between average house prices in Western Australia and average house prices in Australia as a whole has a correlation coefficient, r, equal to 0.83. The coefficient of determination, r^2 is then approximately 0.69. This means that approximately 69% of the variation in the average house price in Western Australia can be explained by the variation in the average house price in Australia as a whole.

Using the correlation coefficient, r, we might be tempted to interpret a linear association for which r = 0.8 as being twice as strong as an association for which r = 0.4. However, the r² values tell us that in the r = 0.8 association 64% of the variation in one variable can be explained by the association between the variables whilst in the r = 0.4 association just 16% is explained. Hence, in terms of the explained variation, the r = 0.8 association is four times the strength of the r = 0.4 association.

EXAMPLE 2

Draw a scattergraph for the data shown below, determine r and r², giving each value correct to two decimal places, and interpret the values obtained.

x	9	42	32	28	45	24	36	14	18	11	15	7
y	49	10	52	33	18	59	21	53	62	95	82	82

Solution

The scattergraph is shown on the right.

Using a calculator,   r   =   −0.84   (to 2 decimal places),

and   r²   =   0.71   (to 2 decimal places).

The value of r informs us that x and y have a strong negative linear association.

The value of r² informs us that 71% of the variation in the variables is explained by the linear association between them.

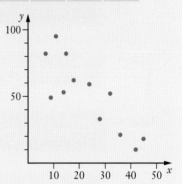

Exercise 2C

1 For the bivariate data:

x	44	40	66	30	76	59	66	31	50	73	57	35
y	73	78	67	81	55	65	60	57	70	60	72	82

Determine the least squares regression line $y = ax + b$ and the correlation coefficient for the data with the outlier   **a**   included,

                    b   excluded.

2 a Display the following bivariate data as a scatterplot with the variable p on the horizontal axis and the variable q on the vertical axis.

p	35	60	20	47	27	61	53	15	70	79	41	56	32	24	43	30	12	62	47	80
q	44	55	34	45	38	69	51	34	43	49	43	54	40	35	47	36	30	52	50	46

b Crop the data so that you are considering the sixteen points that are more suited to linear regression and determine the least squares regression line equation $q = ap + b$ for this cropped data.

c Use your regression line to predict a value for q for a p value of 38.

d Why would it be unwise to use the regression line from part **b** to predict the q value for a p value of 75?

3 The tables shown below give the x-values and the linear regression residuals $(y - \hat{y})$ for six sets of bivariate data. ($\hat{y}$ being the y value predicted by the regression line.)

Following the tables the scattergraphs for these same six sets of bivariate data are shown.

Match each table to its associated graph.

Table 1

x	1	2	3	4	5	6	7	8	9	10
Residuals $(y - \hat{y})$	6.0	2.1	0.2	−3.7	−3.6	−5.4	−3.3	−1.2	0.9	8.0

Table 2

x	1	2	3	4	5	6	7	8	9	10
Residuals $(y - \hat{y})$	0	0	0	0	0	0	0	0	0	0

Table 3

x	1	2	3	4	5	6	7	8	9	10
Residuals $(y - \hat{y})$	−3.4	1.3	2.1	−0.2	1.5	2.3	−1.0	−3.3	−0.5	1.2

Table 4

x	1	2	3	4	5	6	7	8	9	10
Residuals $(y - \hat{y})$	15.8	4.9	−4.0	−8.9	−8.8	−9.6	−6.5	−1.4	4.7	13.8

Table 5

x	1	2	3	4	5	6	7	8	9	10
Residuals $(y - \hat{y})$	−5.2	3.4	8.0	10.7	0.3	−14.1	−12.5	−4.8	1.8	12.4

Table 6

x	1	2	3	4	5	6	7	8	9	10
Residuals $(y - \hat{y})$	4.0	−4.2	−2.3	2.5	−1.7	2.1	0.9	−1.3	−1.4	1.4

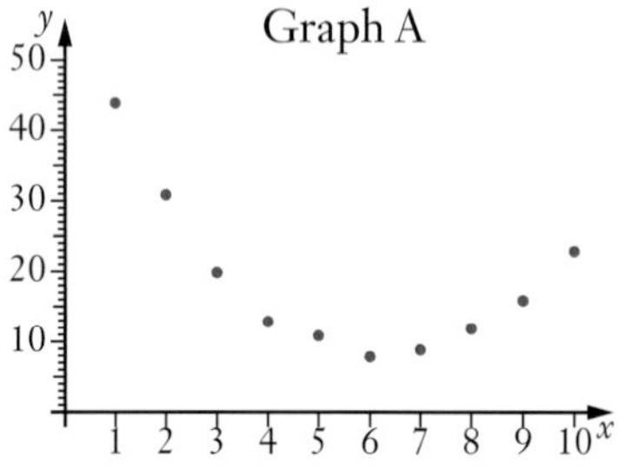

Graph A

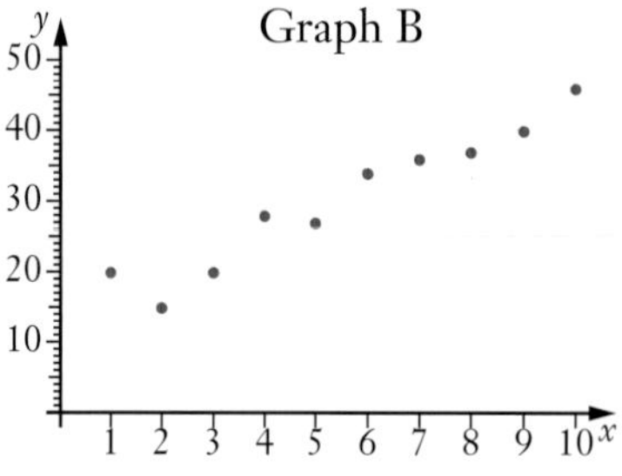

Graph B

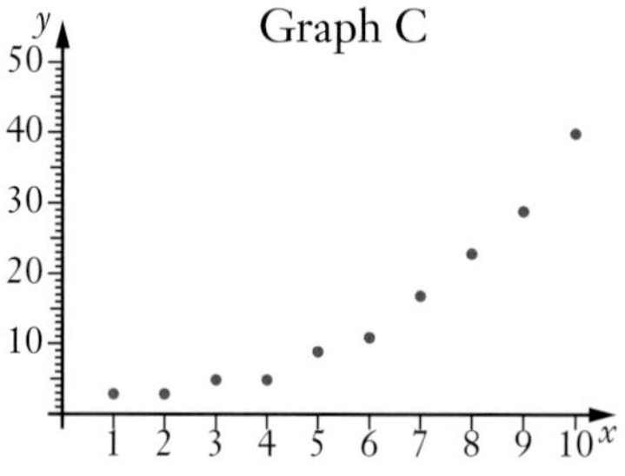

Graph C

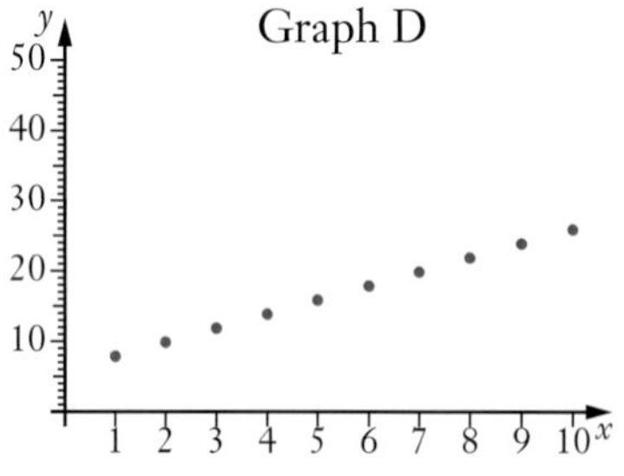

Graph D

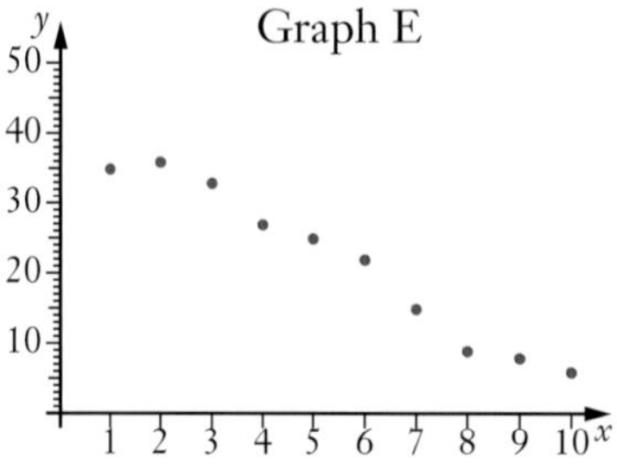

Graph E

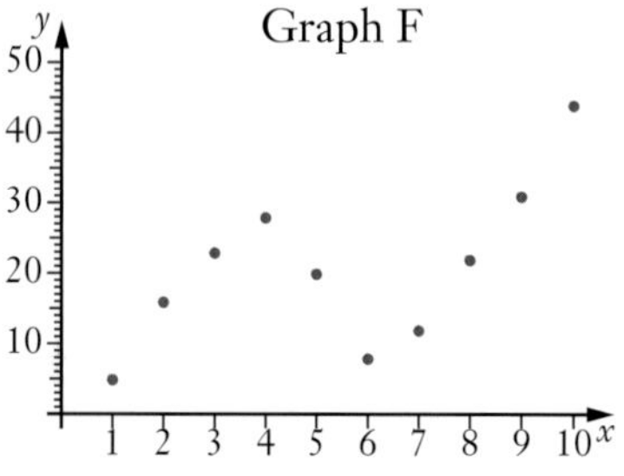

Graph F

ISBN 9780170394994

4 An investigation into heights and weights of men of a particular age found that the correlation between height and weight had a correlation coefficient of 0.802.

According to this result what percentage of the variation in the weight of men of this age can be explained by the variation in their heights?

5 A survey into the foot length of a number of primary school children and the reading age of these children showed a correlation between these two variables with a correlation coefficient, r, equal to 0.73.

According to this result what percentage of the variation in the reading age could be accounted for by the variation in foot length?

Does this mean that increasing the length of a person's foot causes their reading age to increase? Discuss.

6 Draw a scattergraph for the data shown below, determine r and r², giving each value correct to two decimal places, and interpret the values obtained.

x	8	21	20	19	40	53	61	55	72	69
y	10	11	20	37	29	8	24	45	32	49

7 A study of the number of units of fertiliser initially added to the soil mix in seed trays in which seeds of a particular type are to be sown, and the average height of the seedlings in each tray six weeks after sowing, yielded the following results:

Units of fertiliser added initially	0	1	2	3	4	5	6	7	8
Mean height of seedling after six weeks (in millimetres)	73	84	81	93	92	98	105	114	112

View the scattergraph of these results on a graphic calculator or computer and, if linear regression seems appropriate, determine what percentage of the variation in the height of the seedlings can be accounted for by the variation in the number of units of fertiliser added to the seed tray prior to planting.

8 Scientists monitor the concentration in the levels of a pollutant existing in a lake, following an industrial accident and the ongoing clean up.

The pollutant level, P parts per million, present t days after the monitoring commenced is as follows:

t	0	1	2	3	4	5	6	7	8	9	10
P	500	430	314	286	234	185	155	140	118	99	71

a The least squares regression line for this data is $P \approx 431.45 - 40.25t$.

Use this rule to complete the following table of residuals, to the nearest unit and draw the residual plot.

(Residual = observed value – value predicted by regression line).

t days	0	1	2	3	4	5	6	7	8	9	10
Residual											

b Explain why these residuals suggest that *linear* regression may not be the most appropriate.

9 Measurements of two variables, h and r, are recorded by scientists investigating possible links between these two variables.

The recorded measurements are shown in the following table.

h	1	2	3	4	5	6	7	8	9	10
r	15	85	195	345	540	770	1045	1360	1715	2110

a The least squares regression line for this data is $r \approx 233h - 463$.

Use this rule to complete the following table of residuals, to the nearest unit.

(Residual = observed value – value predicted by regression line).

h	1	2	3	4	5	6	7	8	9	10
Residual										

b Explain why these residuals suggest that *linear* regression may not be the most appropriate.

Extension: Quadratic regression

c Suppose instead that r and h are thought to be related by a rule that is of the form

$$r = ah^2 + bh + c$$

This is now in the form of a *quadratic equation*, not a linear equation.

Using a calculator that can perform quadratic regression, determine values for a, b and c.

d Use the quadratic rule from **c** to predict r when $h = 15$ commenting on the reliability of your prediction.

ISBN 9780170394994

ANSCOMBE'S QUARTET*

For *each* of the four data sets shown below use the statistical facilities of your calculator to determine the following, correct to two decimal places.

- The mean of the x scores.

- The standard deviation of the x scores.

- The mean of the y scores.

- The standard deviation of the y scores.

- The coefficient of linear correlation, r_{xy}.

- The equation of the y on x linear regression line.

Present your results neatly and comment on anything you notice.

As you carry out the above also view the scattergraph for each data set and comment on the suitability of using linear regression techniques for each set of data.

Data Set One

x	10.0	8.0	13.0	9.0	11.0	14.0	6.0	4.0	12.0	7.0	5.0
y	8.04	6.95	7.58	8.81	8.33	9.96	7.24	4.26	10.84	4.82	5.68

Data Set Two

x	10.0	8.0	13.0	9.0	11.0	14.0	6.0	4.0	12.0	7.0	5.0
y	9.14	8.14	8.74	8.77	9.26	8.10	6.13	3.10	9.13	7.26	4.74

Data Set Three

x	10.0	8.0	13.0	9.0	11.0	14.0	6.0	4.0	12.0	7.0	5.0
y	7.46	6.77	12.74	7.11	7.81	8.84	6.08	5.39	8.15	6.42	5.73

Data Set Four

x	8.0	8.0	8.0	8.0	8.0	8.0	8.0	19.0	8.0	8.0	8.0
y	6.58	5.76	7.71	8.84	8.47	7.04	5.25	12.50	5.56	7.91	6.89

*F. J. Anscombe of Yale University presented these four data sets in an article published in *The American Statistician*, Volume 27, 1973. The sets are reproduced here with the permission of *The American Statistical Association*.

Data collection, display and analysis

Collect data to investigate one of the questions listed below (or a similar sort of question of your own) and write a report of your investigations. Your report should include the following:

- The question you decided to investigate.

- What data you collected.

- How you collected the data.

- The data, tabulated and/or graphed as appropriate.

- Your conclusions and reasoning. (Including mention of any improvements that could be made to the investigation were it to be repeated.)

QUESTIONS

- Is there a relationship between the height of parents and the heights of their children?

- Is there a relationship between the age of the intended readership of a children's book and average sentence length?

- If a year 8 student is good at mathematics are they also likely to be good at English?

- Is there a relationship between height and hand span?

- Is the number of cold drinks sold in the school canteen during lunchtime related to the temperature of the day?

Miscellaneous exercise two

This miscellaneous exercise may include questions involving the work of this chapter, the work of any previous chapters, and the ideas mentioned in the Preliminary work section at the beginning of the book.

1 A study of a number of fires occurring in a metropolitan area found that there was a strong positive correlation between the number of firemen attending a fire and the amount of damage done by the fire. With this high positive correlation showing that

more firemen attending,
more damage done

would we be correct to conclude that the firemen were causing the damage? Explain your answer.

Shutterstock.com/Johnny Habell

2 Is there an association between whether or not someone is a regular cyclist and gender?

Let us suppose that to investigate the above question researchers defined what they meant by being a 'regular cyclist' and then asked a number of people, all aged between 18 and 50, if they were regular cyclists. The results gave rise to the following table.

	Male	Female
Regular cyclist	52	19
Not a regular cyclist	132	102

Use these results to suggest an answer to the question posed at the beginning of this question and explain your reasoning.

3 In the course of an experiment scientists record values of two variable quantities, x and y, at a number of times. The paired values recorded were as follows:

x	51.2	52.2	53.1	50.6	52.3	51.6	50.9	52.8	51.7	50.2	50.1
y	31.3	26.5	26.2	36.7	30.7	31.5	34.2	25.1	31.3	36.3	37.8

Construct a scattergraph of these results and, purely by 'eyeballing' your graph, estimate the value of the variable y **a**   when x is 50.8,

          b   when x is 53.2,

and   **c**   when x is 58,

in each case commenting on the likely reliability of your prediction.

4 To pass a particular university unit students have to average at least 50% in three tests. If a student misses one of the tests, and has a valid reason for doing so, an estimated mark can be given. Sixteen students take the unit. Fifteen complete all three tests and student number 12 misses test C due to hospitalisation. The percentages achieved by the fifteen students sitting all three tests are given below:

Student No.	1	2	3	4	5	6	7	8	9	10	11	13	14	15	16
Test A	30	86	36	66	40	44	76	10	33	74	23	53	77	16	55
Test B	60	91	41	57	45	60	70	12	22	80	40	36	71	9	86
Test C	40	76	50	88	70	62	70	40	60	79	26	75	94	53	63

a By comparing correlation coefficients determine which of tests A and B is the better indicator of performance in test C for a linear model.

b Use this 'better indicator' to estimate a test C mark (to nearest 1%) for student 12 given that they scored 59% in test A and 48% in test B.

c Which of the sixteen students do not pass the unit?

5 The scattergraph below is for the marks obtained by 30 students who sat two tests, A and B. The associated boxplots are also shown.

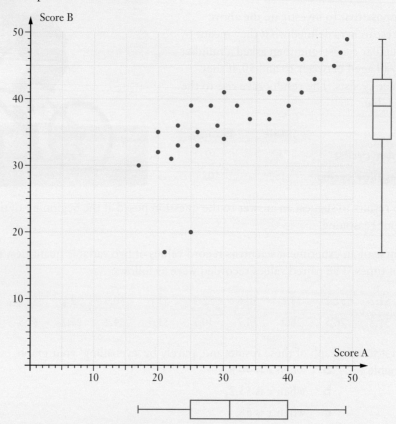

a Compare the location and spread of the marks in the two tests.

b The data is to be used to estimate a test B score for a 31st student who scored 27 in test A but missed test B. Determine an appropriate regression line, explaining your choice, state its equation, make the estimate and comment on the likely validity of your estimate.

In the first half of the 20th century the increasing death rate due to lung cancer in a number of countries was causing concern. Some people suggested that smoking was to blame. Some argued that whilst there might be an association between smoking and lung cancer the association was not necessarily causal.

More recently the debate has expanded to include consideration of 'passive' smoking, the name given to the inhalation of 'second-hand' smoke.

Use the internet to research this correlation/causation debate regarding smoking and lung cancer.

3.

Sequences – recursion

- Terms of sequences
- Recursive rules
- Miscellaneous exercise three

Exercise 3A

Write down the next two entries for each of the following.

1	1,	3,	5,	7,	9,	11,	_____ ,	_____ .
2	5,	10,	20,	40,	80,	160,	_____ ,	_____ .
3	2,	4,	6,	8,	10,	12,	_____ ,	_____ .
4	A,	B,	C,	D,	E,	F,	_____ ,	_____ .
5	A,	D,	G,	J,	M,	P,	_____ ,	_____ .
6	I,	II,	III,	IV,	V,	VI,	_____ ,	_____ .
7	4,	7,	10,	13,	16,	19,	_____ ,	_____ .
8	J,	F,	M,	A,	M,	J,	_____ ,	_____ .
9	1,	1,	2,	3,	5,	8,	_____ ,	_____ .
10	1,	4,	9,	16,	25,	36,	_____ ,	_____ .
11	2,	3,	5,	7,	11,	13,	_____ ,	_____ .
12	1,	3,	6,	10,	15,	21,	_____ ,	_____ .
13	10,	1,	0.1,	0.01,	0.001,	0.0001,	_____ ,	_____ .
14	1,	2,	4,	8,	16,	32,	_____ ,	_____ .
15	1,	3,	9,	27,	81,	243,	_____ ,	_____ .
16	100,	99,	97,	94,	90,	85,	_____ ,	_____ .
17	1234,	2341,	3412,	4123,	1234,	2341,	_____ ,	_____ .
18	7,	18,	29,	40,	51,	62,	_____ ,	_____ .
19	1,	0.5,	0.25,	0.125,	0.0625,	0.03125,	_____ ,	_____ .
20	1,	0.5,	$0.\overline{3}$,	0.25,	0.2,	$0.1\overline{6}$,	_____ ,	_____ .
21	1,	3,	7,	15,	31,	63,	_____ ,	_____ .
22	11,	9,	7,	5,	3,	1,	_____ ,	_____ .
23	32,	48,	72,	108,	162,	243,	_____ ,	_____ .
24	1,	4,	27,	256,	3125,	46656,	_____ ,	_____ .

Terms of sequences

How did you get on with continuing the **sequences** on the previous page?

> A **sequence** is a set of items belonging in a certain order, according to some rule.
> Knowing a particular item and the rule that is involved, the next item can be determined.

The previous page involved sequences following many different rules.

Some sequences were probably familiar to you from everyday life:

For example:

| A, | B, | C, | D, | E, | F, | ... | the letters of the alphabet. |
| J, | F, | M, | A, | M, | J, | ... | the months of the year. |

Others may have been familiar types of number:

For example:

| 2, | 3, | 5, | 7, | 11, | 13, | ... | the prime numbers. |
| 1, | 4, | 9, | 16, | 25, | 36, | ... | the square numbers. |

Some of the sequences continued by repeatedly adding a number:

For example:

4,	7,	10,	13,	16,	19,	...	adding 3 each time.
7,	18,	29,	40,	51,	62,	...	adding 11 each time.
11,	9,	7,	5,	3,	1,	...	adding −2 each time.

Others continued by repeatedly multiplying by a number:

For example:

5,	10,	20,	40,	80,	160,	...	× by 2 each time.
1,	3,	9,	27,	81,	243,	...	× by 3 each time.
1,	0.5,	0.25,	0.125,	0.0625,	0.03125,	...	× by 0.5 each time.

Others may have followed other rules.

In this unit we will concentrate on *number* sequences and we call each number in the sequence a **term** of the sequence.

Thus in the number sequence:

$$4, \qquad 7, \qquad 10, \qquad 13, \qquad 16, \qquad 19, \qquad ...$$

the first term is 4, the second term is 7, the third term is 10, the fourth term is 13, etc.

Writing T_1 for the first term, T_2 for the second term and so on we write this as:

$$T_1 = 4, \qquad T_2 = 7, \qquad T_3 = 10, \qquad T_4 = 13, \qquad T_5 = 16 \qquad T_6 = 19, \qquad ...$$

EXAMPLE 1

For the sequence 5, 11, 23, 47, 95, 191, 383, ...   determine

a T_3 **b** T_5 **c** $T_3 + T_5$ **d** $3T_2$ **e** $2T_3$ **f** T_8

Solution

a T_3 is the 3rd term in the sequence. Thus $T_3 = 23$.

b T_5 is the 5th term in the sequence. Thus $T_5 = 95$.

c $\begin{aligned} T_3 + T_5 &= 23 + 95 \\ &= 118 \end{aligned}$ **d** $\begin{aligned} 3T_2 &= 3(11) \\ &= 33 \end{aligned}$ **e** $\begin{aligned} 2T_3 &= 2(23) \\ &= 46 \end{aligned}$

f   Noticing that each term is obtained by multiplying the previous term by 2 and then adding 1, it follows that

$$\begin{aligned} T_8 &= 2(T_7) + 1 \\ &= 2(383) + 1 \\ &= 767. \end{aligned}$$

Note: Whilst it is usual to use the letter T for the terms of a sequence, other letters are sometimes used as we shall see in the next exercise.

Exercise 3B

For the sequence $5, 8, 11, 14, 17, 20, 23, 26, 29, \ldots$ determine

1 T_4
2 T_5
3 $T_4 + T_5$
4 $3T_4$
5 $4T_3$
6 $T_3 + T_5$
7 $3(T_1 + T_2)$
8 $3T_1 + T_2$
9 T_{10}
10 $T_2 - T_1$
11 $T_3 - T_2$
12 $T_4 - T_3$

For the sequence $3, 7, 15, 31, 63, 127, 255, \ldots$ determine

13 T_2
14 T_7
15 $T_2 + T_7$
16 $3T_2$
17 $2T_3$
18 $T_4 + 2T_5$
19 $2T_3 - T_2$
20 $2(T_3 - T_2)$
21 T_8
22 $T_2 - T_1$
23 $T_3 - T_2$
24 $T_4 - T_3$

For the sequence $2, 5, 10, 17, 26, 37, 50, \ldots$ determine

25 T_5
26 $3T_2$
27 $T_1 + T_2 + T_3$
28 $T_2 + T_3 + T_4$
29 $T_2 - T_1$
30 $T_3 - T_2$
31 $T_4 - T_3$
32 T_8

The sequence $1, 1, 2, 3, 5, 8, 13, 21, 34, \ldots$ is known as the Fibonacci sequence. For this sequence we tend to use F_1 for the first term, F_2 for the second term etc. Determine

33 F_1
34 F_4
35 $F_1 + F_2$
36 $F_2 + F_3$
37 $F_3 + F_4$
38 $F_6 + F_7 + F_8$
39 $F_6 + F_7 - F_8$
40 F_{10}

The sequence $1, 3, 4, 7, 11, 18, 29, 47, 76, \ldots$ is known as the Lucas sequence. For this sequence we tend to use L_1 for the first term, L_2 for the second term etc. Determine

41 L_1
42 L_4
43 $L_1 + L_2$
44 $L_2 + L_3$
45 $L_3 + L_4$
46 $L_6 + L_7 + L_8$
47 $L_6 + L_7 - L_8$
48 L_{10}

Recursive rules

Consider the sequence: $5, \quad 8, \quad 11, \quad 14, \quad 17, \quad 20, \quad 23, \quad \ldots$

Each term is obtained by adding 3 to the previous term.

The statement *Each term is obtained by adding 3 to the previous term* tells us how the sequence **recurs**. It is the **recursive rule** for the sequence.

Knowing the recursive rule for a sequence and at least one term, often the first, allows the terms of the sequence to be determined.

EXAMPLE 2

Generate the first six terms of a sequence for which:

$$\begin{cases} \text{The first term of the sequence is 73} \\ \text{Each term is 5 less than the previous term} \end{cases}$$

Solution

The first six terms will be:

$$\begin{array}{rcll} & & 73 & \leftarrow \quad \text{1st term} \\ 73 - 5 & = & 68 & \leftarrow \quad \text{2nd term} \\ 68 - 5 & = & 63 & \leftarrow \quad \text{3rd term} \\ 63 - 5 & = & 58 & \leftarrow \quad \text{4th term} \\ 58 - 5 & = & 53 & \leftarrow \quad \text{5th term} \\ 53 - 5 & = & 48 & \leftarrow \quad \text{6th term} \end{array}$$

In the above example:

$$T_1 = 73, \qquad T_2 = 68, \qquad T_3 = 63, \qquad T_4 = 58, \qquad T_5 = 53, \qquad T_6 = 48.$$

Writing T_n for the nth term it follows that the next term, T_{n+1}, will be $T_n - 5$.

We can then write the recursive rule *each term is 5 less than the previous term* as

$$T_{n+1} \quad = \quad T_n - 5.$$

To be able to determine the terms of the sequence we also need to know one term. Hence the sequence 73, 68, 63, 58, 53, 48, … can be defined by stating:

$$T_{n+1} \quad = \quad T_n - 5, \qquad T_1 \quad = \quad 73.$$

Read the following and check that you agree with the word description of each recursive formula.

Recursive formula	Word description
$T_{n+1} = T_n + 4$	Each term is obtained by adding 4 to the previous term.
$T_{n+1} = T_n - 2$	Each term is obtained by subtracting 2 from the previous term.
$T_{n+1} = 5T_n$	Each term is obtained by multiplying the previous term by 5.
$T_{n+1} = 3T_n + 2$	Each term is obtained by multiplying the previous term by 3 and then adding 2.
$T_{n+2} = T_n + T_{n+1}$	Each term is obtained by adding the two previous terms.

Note: • If the recursive rule $T_{n+1} = T_n - 5$ is written as $T_{n+1} - T_n = -5$ it is called the **difference rule** for the sequence. In this form the rule makes clear the **difference** between any term and the previous one.

• The recursive rule $T_{n+1} = T_n - 5$ could equally well be written in the form

$$T_n = T_{n-1} - 5.$$

Both expressions tell us the same thing, i.e. that each term of the sequence is obtained by subtracting 5 from the previous term. (See example 3.)

EXAMPLE 3

A sequence is defined by $T_n = T_{n-1} + 3$ with $T_1 = 5$.

Determine the first five terms of this sequence.

Solution

The formula	T_n	$=$	$T_{n-1} + 3$	tells us that each term is obtained by adding 3 to the previous term.				

Thus if	T_1	$=$	5	it follows that	T_2	$=$	$5 + 3$	$=$	$8,$
					T_3	$=$	$8 + 3$	$=$	$11,$
					T_4	$=$	$11 + 3$	$=$	$14,$
					T_5	$=$	$14 + 3$	$=$	$17.$

The first five terms are 5, 8, 11, 14, 17.

EXAMPLE 4

A sequence is defined by $T_{n+1} = 2T_n + 3$ with $T_1 = 6$.

Determine the first five terms of this sequence.

Solution

The formula $T_{n+1} = 2T_n + 3$ tells us that each term is obtained by multiplying the previous term by 2 and then adding 3.

Thus if	T_1	$=$	6	it follows that	T_2	$=$	$2 \times 6 + 3$	$=$	$15,$
					T_3	$=$	$2 \times 15 + 3$	$=$	$33,$
					T_4	$=$	$2 \times 33 + 3$	$=$	$69,$
					T_5	$=$	$2 \times 69 + 3$	$=$	$141.$

The first five terms are 6, 15, 33, 69, 141.

EXAMPLE 5

A sequence is defined by $T_n = 0.5T_{n-1}$ with $T_1 = 400$.

Determine the first five terms of this sequence.

Solution

The formula $T_n = 0.5T_{n-1}$ tells us that each term is obtained by multiplying the previous term by 0.5.

Thus if	T_1	$=$	400	it follows that	T_2	$=$	$200,$
					T_3	$=$	$100,$
					T_4	$=$	$50,$
					T_5	$=$	$25.$

The first five terms are 400, 200, 100, 50, 25.

EXAMPLE 6

A sequence is defined by $T_{n+1} = 4T_n - 2$ with $T_1 = 3$.

Determine the first five terms of this sequence.

Solution

The formula $T_{n+1} = 4T_n - 2$ tells us that each term is obtained by multiplying the previous term by 4 and then subtracting 2.

Thus if $T_1 = 3$ it follows that

$$T_2 = 4(3) - 2 = 10,$$
$$T_3 = 4(10) - 2 = 38,$$
$$T_4 = 4(38) - 2 = 150,$$
$$T_5 = 4(150) - 2 = 598.$$

The first five terms are 3, 10, 38, 150, 598.

Alternatively, rather than thinking and interpreting the recursive formula

$$T_{n+1} = 4T_n - 2$$

as meaning multiply the previous term by 4 and then subtract 2, we could simply substitute into the given formula:

To obtain T_2 substitute $n = 1$ into $T_{n+1} = 4T_n - 2$ $\quad T_2 = 4T_1 - 2 = 10$

To obtain T_3 substitute $n = 2$ into $T_{n+1} = 4T_n - 2$ $\quad T_3 = 4T_2 - 2 = 38$

To obtain T_4 substitute $n = 3$ into $T_{n+1} = 4T_n - 2$ $\quad T_4 = 4T_3 - 2 = 150$

To obtain T_5 substitute $n = 4$ into $T_{n+1} = 4T_n - 2$ $\quad T_5 = 4T_4 - 2 = 598$

Note: This substitution method may seem rather formal and tedious but it can be useful if the recursive formula is not easy to understand intuitively. (See example 7).

EXAMPLE 7

A sequence is defined by $T_{n+1} = 2^n T_n$ with $T_1 = 5$.

Determine the first four terms of this sequence.

Solution

With $n = 1$ $\quad T_{n+1} = 2^n T_n$ becomes $\quad T_2 = 2^1 T_1$
$$= 2 \times 5$$
$$= 10$$

With $n = 2$ $\quad T_{n+1} = 2^n T_n$ becomes $\quad T_3 = 2^2 T_2$
$$= 4 \times 10$$
$$= 40$$

With $n = 3$ $\quad T_{n+1} = 2^n T_n$ becomes $\quad T_4 = 2^3 T_3$
$$= 8 \times 40$$
$$= 320$$

The first four terms are 5, 10, 40, 320.

Alternatively the recursive formula could be put into a calculator and the terms of the sequence displayed, as shown on the right for the previous example.

Note that the display on the right also shows the progressive sums (also called partial sums).

$$T_1 = 5$$
$$T_1 + T_2 = 5 + 10 = 15$$
$$T_1 + T_2 + T_3 = 5 + 10 + 40 = 55$$
etc.

Calculators differ in the way they accept and display such information. If you wish to use a calculator in this way make sure that **you** can use **your** calculator to input recursive formulae and to display the terms of a sequence.

Whilst it is usual to give the first term of a sequence with the recursive rule, knowing any term will allow other terms to be determined. For example the statements:

$$\begin{cases} \text{The fourth term of the sequence is 13.} \\ \text{Each term is 3 more than the previous term.} \end{cases}$$

also allows the sequence 4, 7, 10, 13, 16, 19, 22, ... to be determined.

☑	$a_{n+1} = 2^n \cdot a_n$
	$a_1 = 5$
☐	$b_{n+1} = \square$
	$b_1 = 0$
☐	$c_{n+1} = \square$
	$c_1 = 0$

n	a_n	Σa_n
1	5	5
2	10	15
3	40	55
4	320	375
5	5120	5495

EXAMPLE 8

Generate the first six terms of a sequence for which:

$$\begin{cases} \text{The fourth term of the sequence is 47.} \\ \text{Each term is twice the previous term add one.} \end{cases}$$

Solution

The task is to determine T_1 to T_6 given the following information:

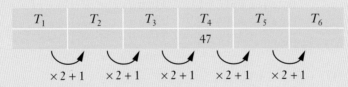

T_1	T_2	T_3	T_4	T_5	T_6
			47		

$\times 2 + 1 \quad \times 2 + 1 \quad \times 2 + 1 \quad \times 2 + 1 \quad \times 2 + 1$

Hence
$$T_5 = 2T_4 + 1 \qquad\qquad T_6 = 2T_5 + 1$$
$$\quad = 95 \qquad\qquad\qquad \quad = 191$$

The third term, and similarly the second and first, could be found intuitively by thinking what number when doubled and 1 is added gives an answer of 47.

Alternatively an algebraic approach is possible, as follows:

$$\begin{aligned} T_4 &= 2T_3 + 1 & T_3 &= 2T_2 + 1 & T_2 &= 2T_1 + 1 \\ 47 &= 2T_3 + 1 & 23 &= 2T_2 + 1 & 11 &= 2T_1 + 1 \\ \therefore \quad 46 &= 2T_3 & 22 &= 2T_2 & 10 &= 2T_1 \\ \text{i.e.} \quad T_3 &= 23 & T_2 &= 11 & T_1 &= 5 \end{aligned}$$

The first six terms of the sequence are 5, 11, 23, 47, 95, 191.

3. Sequences – recursion ●●● ○○○

EXAMPLE 9

State T_1 and the recursive formula for the sequence 98, 91, 84, 77, 70, ...

Solution

By inspection each term in the sequence is 7 less than the term before it.

Thus $T_{n+1} = T_n - 7$ with $T_1 = 98$.

EXAMPLE 10

A sequence has a first term of 240 and each term after the first is equal to the previous term increased by 10%.

State T_1 to T_4 and the recursive formula for the sequence.

Solution

Each term after the first will be the previous term multiplied by 1.1.

Thus $T_1 = 240$, $T_2 = 264$, $T_3 = 290.4$, $T_4 = 319.44$ and $T_{n+1} = 1.1 \times T_n$.

EXAMPLE 11

Given that the sequence 3, 17, 73, 297, 1193, ... has a recursive formula that is of the form $T_{n+1} = aT_n + b$ determine a, b and T_6.

Solution

The values of a and b can be obtained by thoughtful trial and adjustment:

The given terms are all integers which suggests that a and b are integers.

With $T_1 = 3$ and $T_2 = 17$,   try $a = 1$ and $b = 14$.   Then T_3 = $1(17) + 14$ $\neq$ 73

    Try $a = 2$ and $b = 11$.   Then T_3 = $2(17) + 11$ $\neq$ 73

    Try $a = 3$ and $b = 8$.   Then T_3 = $3(17) + 8$ $\neq$ 73

    Try $a = 4$ and $b = 5$.   Then T_3 = $4(17) + 5$ = 73 ✓

    and T_4 = $4(73) + 5$ = 297 ✓

Hence T_6 = $4T_5 + 5$ = $4(1193) + 5$

      = 4777

The required values are $a = 4$, $b = 5$ and $T_6 = 4777$.

Alternatively, using an algebraic approach:

From $T_{n+1} = aT_n + b$ we know that each term is obtained by multiplying the previous term by a and then adding b.

Thus with T_1 = 3   and T_2 = 17   it follows that   17 = $3a + b$,   ← [1]

and with T_2 = 17   and T_3 = 73   it follows that   73 = $17a + b$.   ← [2]

Solving equations [1] and [2] simultaneously gives $a = 4$ and $b = 5$ as before.

Exercise 3C

1 For each description, find an appropriate recursive formula from the list.

Descriptions

a: Each term is obtained by adding 5 to the previous term.

b: Each term is obtained by subtracting 5 from the previous term.

c: Each term is obtained by multiplying the previous term by 5.

d: Each term is obtained by multiplying the previous term by 0.5.

e: Each term is obtained by adding the two previous terms.

f: Each term is obtained by multiplying the previous term by 2 and then subtracting 5.

g: Each term is obtained by adding the three previous terms.

h: Each term is obtained by dividing the previous term by 2.

i: Each term is the product of the previous three terms.

Recursive formulae

A: $T_{n+1} = T_n - 5$ B: $T_{n+1} = 5T_n$

C: $T_{n+1} = (T_n)^5$ D: $T_{n+1} = 0.5(T_n)$

E: $T_{n+2} = T_n - T_{n+1}$ F: $T_{n+3} = (T_n)(T_{n+1})(T_{n+2})$

G: $T_{n+1} = T_n + T_{n-1}$ H: $T_{n+3} = T_n + T_{n+1} + T_{n+2}$

I: $T_{n+1} = T_n + 5$ J: $T_{n+1} = 5T_n + 2$

K: $T_{n+1} = 2T_n - 5$ L: $T_{n+1} = 5T_n - 2$

2 Determine the first six terms of a sequence for which:

$$\begin{cases} \text{The first term of the sequence is 11.} \\ \text{Each term is five more than the previous term.} \end{cases}$$

3 Determine the first six terms of a sequence for which:

$$\begin{cases} \text{The first term of the sequence is 4.} \\ \text{Each term is the previous term doubled and then take away 5.} \end{cases}$$

4 A sequence is defined by $T_n = T_{n-1} + 4$ with $T_1 = 3$. Determine the first five terms of this sequence.

5 A sequence is defined by $T_{n+1} = T_n + 3$ with $T_1 = 1$. Determine the first five terms of this sequence.

6 A sequence is defined by $T_{n+1} = 2T_n + 1$ with $T_1 = 2$. Determine the first five terms of this sequence.

7 A sequence is defined by $T_{n+1} = 3T_n - 1$ with $T_1 = 2$. Determine the first five terms of this sequence.

8 A sequence is defined by $T_n = 5T_{n-1} - 2$ with $T_1 = 1$. Determine the first five terms of this sequence.

9 A sequence has a first term of 128 and each term after the first is equal to the previous term increased by 25%.

State T_1 to T_4 and the recursive formula for the sequence.

10 A sequence has a first term of 1000 and each term after the first is equal to the previous term increased by 20%.

State T_1 to T_4 and the recursive formula for the sequence.

11 A sequence has a first term of 2000 and each term after the first is equal to the previous term decreased by 15%.

State T_1 to T_4 and the recursive formula for the sequence.

12 State T_1 and a recursive formula for the sequence 5, 9, 13, 17, 21, ...

13 State T_1 and a recursive formula for the sequence 26, 23, 20, 17, 14, ...

14 State T_1 and a recursive formula for the sequence 6, 13, 27, 55, 111, ...

15 A sequence has a recursive formula of the form $T_n = 2T_{n-1} + a$, with $T_1 = 3$.
If $T_2 = 10$ find the value of a and determine T_3.

16 Determine the first six terms of a sequence for which:

$$\begin{cases} \text{The second term of the sequence is 23.} \\ \text{Each term is eight more than the previous term.} \end{cases}$$

17 Determine the first six terms of a sequence for which:

$$\begin{cases} \text{The third term of the sequence is 74.} \\ \text{Each term is three times the previous term take 4.} \end{cases}$$

18 Determine the first six terms of a sequence for which:

$$\begin{cases} \text{The fourth term of the sequence is 89.} \\ \text{Each term is twice the previous term take 1.} \end{cases}$$

19 A sequence is defined by $T_{n+2} = T_n + T_{n+1}$ with $T_1 = 1$ and $T_2 = 4$. Determine the first five terms of this sequence.

20 A sequence is defined by $T_n = T_{n-2} + T_{n-1}$ with $T_1 = T_2 = 7$. Determine the first five terms of this sequence.

21 A sequence is defined by $T_{n+2} = 2T_n - T_{n+1}$ with $T_1 = 3$ and $T_2 = 4$. Determine the first five terms of this sequence.

ISBN 9780170394994

22 A sequence is defined by $T_n = T_{n-3} + T_{n-2} + T_{n-1}$ with $T_1 = 1$, $T_2 = 2$ and $T_3 = 3$.
Determine the first six terms of this sequence.

23 Given that the sequence $1, 8, 43, 218, 1093, \ldots$ has a recursive formula that is of the form
$T_{n+1} = aT_n + b$ determine a, b and T_6.

24 Given that the sequence $5, 7, 11, 19, 35, \ldots$ has a recursive formula that is of the form
$T_{n+1} = aT_n + b$ determine a, b and T_6.

25 Sequence A has terms $A_1, A_2, A_3, \ldots$ and sequence B has terms $B_1, B_2, B_3, \ldots$.
Given that $A_1 = 5$, $A_n = A_{n-1} + 3$, $B_1 = 64$, and $B_n = 0.75B_{n-1}$, determine the first five terms
of sequence C given that $C_n = A_n + B_n$.

26 A sequence has the recursive formula $T_{n+1} = (-1)^n 2T_n$, with $T_1 = 1$.
Determine the first five terms of this sequence.

27 A sequence has the recursive formula $T_{n+1} = (-1)^{n+1} 2T_n$, with $T_1 = 3$.
Determine the first five terms of this sequence.

28 A sequence has the recursive formula $T_{n+1} = (n+2)^n T_n$, with $T_1 = 3$.
Determine the first five terms of this sequence.

29 Stacking Cans

a Copy and complete the table shown below.

Stacking cans					
T_1	T_2	T_3	T_4	T_5	T_6
1	3	6	10		

b Which of the following recursive formulae give the same first six terms as you obtained in the
table above?

I: $T_{n+1} = T_n + 2$, $T_1 = 1$.

II: $T_n = T_{n-1} + 2$, $T_1 = 1$.

III: $T_n = 2T_{n-1}$, $T_1 = 1$.

IV: $T_n = T_{n-1} + n$, $T_1 = 1$.

Spreadsheets

Computer spreadsheets can be very useful for generating and displaying the terms of a sequence.

Create the following table on a spreadsheet and, by following each pattern down, complete cells A6, A7, A8, B6, B7, B8, C6, C7, C8.

	A	B	C	D
1	1	6	5	
2	2	7	10	
3	3	8	15	
4	4	9	20	
5	5	10	25	
6				
7				
8				

1 The entry that is to go in cell D1 is to be given by the rule:

$$D1 = A1*B1 - C1 \quad \text{(where * represents multiply).}$$

Similarly $\quad D N = A N*B N - C N$

Complete cells D1 to D8 on your spreadsheet.

2 The numbers in cells D1 to D8 are the first 8 _ _ _ _ _ _ numbers. (What word goes in the space?)

3 If columns A, B and C were continued indefinitely, and column D was filled according to the rule in part **1** above, would column D continue to give the type of numbers mentioned in **2** above? Prove it.

Miscellaneous exercise three

This miscellaneous exercise may include questions involving the work of this chapter, the work of any previous chapters, and the ideas mentioned in the Preliminary work section at the beginning of the book.

1 Write down the next two entries for each of the following sequences.

a	1,	4,	13,	40,	121,	364,	_____,	_____.
b	1,	8,	27,	64,	125,	216,	_____,	_____.
c	1000,	800,	640,	512,	409.6,	327.68,	_____,	_____.
d	1000,	800,	600,	400,	200,	0,	_____,	_____.
e	$\frac{1}{8}$,	$\frac{1}{4}$,	$\frac{3}{8}$,	$\frac{1}{2}$,	$\frac{5}{8}$,	$\frac{3}{4}$,	_____,	_____.

2 A sequence is defined by $T_{n+1} = 7T_n - 5$ with $T_1 = 1$.

Determine the first four terms of this sequence.

3 A sequence is defined by $T_n = 0.5T_{n-1} - 2$ with $T_1 = 120$.

Determine the first four terms of this sequence.

4 At various places in this book we have considered the food consumption/heart disease death rate situation. We found that with y as the deaths due to heart disease, in deaths per 100 000 of population, and x as the foodstuff consumption, in kilograms per person per year, the association could be quite well modelled by the least squares regression line.

The data collected suggested that the situation could be well modelled using a line of best fit with equation

$$y = 3.1x + 90.9.$$

Interpret the numbers 3.1 and 90.9 in this situation.

5 A sequence has a recursive formula of the form $T_n = aT_{n-1} + 3$, with $T_1 = 2$.

Given that $T_2 = 25$ find the value of a and determine T_3.

6 Sequence D has terms $D_1, D_2, D_3, \ldots$ and sequence E has terms $E_1, E_2, E_3, \ldots$.

Given that $D_1 = 2$, $D_n = D_{n-1} + 4$, $E_1 = 10$, and $E_n = E_{n-1} - 3$, determine the first five terms of sequence F given that $F_n = (D_n)(E_n)$.

7 A scientific experiment involved a student collecting bivariate data for the two variables x and L.

The tabulated results were as follows:

x	0.5	1.0	1.5	2.0	2.5	3.0	3.5	4.0
L	6.9	7.4	9.2	13.4	19.5	28.3	41.3	58.1

Applying linear regression techniques to the data the student found that the correlation coefficient was 0.93 and the least squares regression line had equation:

$$L = 14.1x - 8.7.$$

The student concluded that linear modelling of the situation was the most appropriate and that $L = 14.1x - 8.7$ was the most appropriate equation to use to represent the relationship algebraically.

Comment on this conclusion.

8 The table below gives the *Number of pupils per teacher* (in first year of schooling) and the *Adult Literacy Rate* (percentage of adult population who are literate) for 22 countries.

Country	A	B	C	D	E	F	G	H	I	J	K
Pupils per teacher	22	28	27	41	26	18	25	51	37	25	19
Literacy rate (%)	95	54	92	51	88	99	65	54	98	60	85

Country	L	M	N	O	P	Q	R	S	T	U	V
Pupils per teacher	36	19	46	16	31	23	63	27	31	67	36
Literacy rate (%)	67	95	48	99	35	81	35	90	87	30	96

a Display the data as a scattergraph on either a calculator display or spreadsheet facility with the pupils per teacher on the horizontal axis and literacy rate on the vertical axis.

b On the basis of this data could we conclude that a high literacy rate is due to a low number of pupils per teacher?

c Estimate the literacy rate for a country with 55 pupils per teacher.

9 a In the section about the line of best fit the equation of the least squares regression line of y on x was expressed as:

$$y - \bar{y} = \frac{s_{xy}}{s_x^2}(x - \bar{x})$$

Prove that this line must pass through the point $(\bar{x}, \bar{y})$.

b Why is the line of best fit referred to as the '**least squares**' regression line?

10 In 2013, in Western Australia, the numbers of year 12 students of each gender achieving the various grades in the top unit of Mathematics (non-specialist) and the top unit in Biology, were as follows:

Mathematics:

	A	B	C	D	E	Total
Male	616	558	867	253	68	2362
Female	415	383	569	140	25	1532

School Curriculum and Standards Authority. 2013 Year 12 Student Achievement Data.
Reproduced with permission from the Government of Western Australia, School Curriculum and Standards Authority. © School Curriculum and Standards Authority, Government of Western Australia

Biology:

	A	B	C	D	E	Total
Male	77	144	306	72	7	606
Female	214	313	415	71	14	1027

School Curriculum and Standards Authority. 2013 Year 12 Student Achievement Data.
Reproduced with permission from the Government of Western Australia, School Curriculum and Standards Authority. © School Curriculum and Standards Authority, Government of Western Australia

Analyse, comment, report.

ISBN 9780170394994

4.

Sequences – some specific types

- Arithmetic sequences
- Geometric sequences
- Jumping to later terms of arithmetic and geometric sequences
- Growth and decay
- First-order linear recurrence relations
- Miscellaneous exercise four

Suppose a business buys an item of machinery costing $250 000. Whilst this purchase will mean that the money the business has in the bank will decrease by $250 000, the business now has an asset worth $250 000. Hence the business has not lost $250 000, it has just re-allocated it from money in the bank to machinery. However, one year later the machine will no longer be new. Its value is likely to have depreciated. The company no longer has an asset worth $250 000. For accounting purposes the business will need to be able to quote a value for the machine but how much will it be worth after one year, two years, three years, etc.? Consider each of the following situations:

Situation One   Depreciation as a fixed annual amount.
(Flat rate, or straight line, method of depreciation.)

Let us suppose that the machine, initially worth $250 000, depreciates by the same amount each year and that after five years it is only worth its 'scrap value' of $25 000.

Copy and complete:

Initial value	$250 000
Value after 1 year	
Value after 2 years	
Value after 3 years	
Value after 4 years	
Value after 5 years	$25 000

Situation Two   Depreciation by the number of units produced.
(Unit cost, or units of production, method of depreciation.)

Suppose instead that the machine is expected to produce 600 000 units, after which it will only have scrap value of $25 000, depreciation occurring at a fixed rate per unit produced.

Copy and complete:

Initial value	$250 000
Value after 100 000 units produced	
Value after 200 000 units produced	
Value after 300 000 units produced	
Value after 400 000 units produced	
Value after 500 000 units produced	
Value after 600 000 units produced	$25 000

Situation Three   Depreciation by a percentage of the declining value.
(Declining balance, or reducing balance, method of depreciation.)

Suppose now that the value of the machine at the end of each year is 80% of its value at the beginning of that year, i.e. a depreciation rate of 20%.

Copy and complete:

Initial value	$250 000
Value after 1 year	
Value after 2 years	
Value after 3 years	

Situation one and situation two on the previous page each involved the amount reducing at a fixed amount, either per year or per 100 000 units.

Situation three involved the amount reducing by virtue of repeated multiplication by a number, in this case 0.8. This number being between 0 and 1 producing the sequence of amounts that were decreasing.

Sequences of numbers for which each term, after the first, is obtained by adding a constant amount to the previous term are called **arithmetic sequences**.

Sequences of numbers for which each term, after the first, is obtained by multiplying the previous term by a constant amount are called **geometric sequences**.

Arithmetic sequences

Notice that in the table of values on the right, as the x-values increase by 1 the y-values increase by 2. As we would expect from this constant first difference pattern of 2, graphing gives points that lie in a straight line and the gradient of the line is 2.

x	1	2	3	4	5	6
y	1	3	5	7	9	11

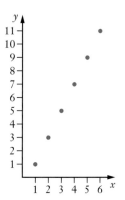

As mentioned above, sequences like 1, 3, 5, 7, 9, … in which each term is obtained from the previous term by the addition of some constant number are said to be

arithmetic sequences.

These are also known as   Arithmetic Progressions
or simply   APs.

For example, the AP: 1,   2,   3,   4,   5,   6,   …

In this case we say that the **first term** is 1 and the **common difference** is 1.

Similarly for the AP: 1,   3,   5,   7,   9,   11,   …
we say that the first term is 1 and the common difference is 2.

For   7,   11,   15,   19,   23,   27,   …
the first term is 7 and the common difference is 4.

For   67,   62,   57,   52,   47,   42,   …
the first term is 67 and the common difference is −5.

Thus all arithmetic sequences are of the form:

$$a, \quad a+d, \quad a+2d, \quad a+3d, \quad a+4d, \quad a+5d, \quad a+6d, \quad …$$

In this general form we have a first term of 'a' and common difference 'd'.

Hence, to define this general arithmetic sequence recursively:

$$T_{n+1} = T_n + d, \qquad\qquad T_1 = a.$$

(Sometimes written with lower case t:

$$t_{n+1} = t_n + d, \qquad\qquad t_1 = a.)$$

EXAMPLE 1

An arithmetic sequence is such that $T_{n+1} = T_n + 5$ and the first term, T_1, is 7.

Find the first four terms of the sequence.

Solution

The recursive definition informs us that each term is the previous term add 5.

Hence if $\quad T_1 \;=\; 7 \quad$ it follows that $\quad$
$$T_2 \;=\; 7 + 5$$
$$ \;=\; 12$$
$$T_3 \;=\; 12 + 5$$
$$ \;=\; 17$$
$$T_4 \;=\; 17 + 5$$
$$ \;=\; 22$$

The first four terms of the sequence are 7, 12, 17, 22.

EXAMPLE 2

For each of the following sequences state whether the sequence is an AP or not and, for those that are, state the first term, the common difference and a recursive formula.

a   13,   18,   23,   28,   33,   38,   ...,

b   3,   6,   12,   24,   48,   96,   ...,

c   90,   79,   68,   57,   46,   35,   ...,

Solution

a   Each term is 5 more than the previous term.

Thus the sequence *is* an arithmetic progression with first term $\quad=\quad 13$

the common difference $\quad=\quad 5$

and $\quad T_{n+1} \;=\; T_n + 5$

b   The terms do not have a common difference.

Thus the sequence is *not* an arithmetic progression.

c   Each term is 11 less than the previous term.

Thus the sequence *is* an arithmetic progression with first term $\quad=\quad 90$

the common difference $\quad=\quad -11$

and $\quad T_{n+1} \;=\; T_n - 11$

EXAMPLE 3

A sequence is defined by $T_n = 2T_{n-1} + 1$ with $T_1 = 3$. Determine the first five terms of the sequence and hence determine whether the sequence is an arithmetic sequence.

Solution

The formula $\quad T_n \;=\; 2T_{n-1} + 1 \quad$ tells us that each term is obtained by multiplying the previous term by 2 and then adding 1.

Thus if $\quad T_1 \;=\; 3 \quad$ it follows that
$$
\begin{aligned}
T_2 &= 2(3) + 1 &&= 7, \\
T_3 &= 2(7) + 1 &&= 15, \\
T_4 &= 2(15) + 1 &&= 31, \\
T_5 &= 2(31) + 1 &&= 63.
\end{aligned}
$$

The first five terms are $3, 7, 15, 31, 63$.

These terms do *not* have a common difference.

The sequence is *not* an arithmetic sequence.

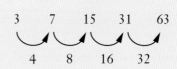

Alternatively, and as mentioned in the previous chapter, a calculator or computer spreadsheet can be used to display the terms of a sequence (and the partial sums if so desired) once an appropriate recursive rule and first term have been entered.

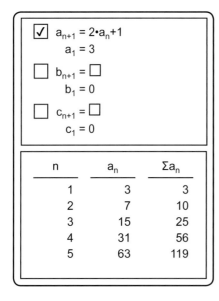

	A	B	C	D
1	3	3		
2	7	10		
3	15	25		
4	31	56		
5	63	119		
6	127	246		
7	255	501		
8	511	1012		
9	1023	2035		
10	2047	4082		

=2*A3+1

=Sum(A$1:A6)

What does the inclusion of the $ symbol do?

Create the spreadsheet yourself and use the 'fill down' ability when creating it.

Geometric sequences

Notice that in the table of values on the right, as the
x-values increase by 1 the y-values multiply by 2.
Graphing the values gives the shape shown on the right which you
may recognise as the characteristic shape of an **exponential** graph.

x	1	2	3	4	5	6
y	3	6	12	24	48	96

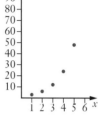

As mentioned previously, sequences that progress by each term being obtained
by multiplying the previous term by a constant number are said to be

geometric sequences.

These are also known as *Geometric Progressions*
or simply *GPs.*

For example the GP 5, 15, 45, 135, 405, 1215, ...

In this case we say that the **first term** is 5 and the **common ratio** is 3.

Similarly for the GP: 0.5, 1, 2, 4, 8, 16, ...
we say that the first term is 0.5 and the common ratio is 2.

For 1000, 100, 10, 1, 0.1, 0.01, ...
the first term is 1000 and the common ratio is 0.1.

For 64, 96, 144, 216, 324, 486, ...
the first term is 64 and the common ratio is 1.5.

Thus all GPs are of the form:

$$a, \quad ar, \quad ar^2, \quad ar^3, \quad ar^4, \quad ar^5, \quad ar^6, \quad \ldots$$

In this general form we have a first term of 'a' and common ratio 'r'.

Using recursive notation we have $T_{n+1} = r \times T_n$ with $T_1 = a$, (or $t_{n+1} = rt_n, t_1 = a$).

EXAMPLE 4

For each of the following sequences state whether the sequence is a geometric sequence or not and,
for those that are, state the first term, the common ratio and a recursive formula.

a 3, 6, 12, 24, 48, 96, ... ,
b 128, 96, 72, 54, 40.5, 30.375, ... ,
c 4, 9, 14, 19, 24, 29, ... ,

Solution

a $\dfrac{6}{3} = 2$, $\dfrac{12}{6} = 2$, $\dfrac{24}{12} = 2$, $\dfrac{48}{24} = 2$, $\dfrac{96}{48} = 2$.

Each term is the previous term multiplied by 2.

Thus the sequence *is* a geometric sequence with first term = 3
the common ratio = 2
and T_{n+1} = $2T_n$
(or T_n = $2T_{n-1}$)

---

b The given sequence is: 128,   96,   72,   54,   40.5,   30.375,   … ,

$\dfrac{96}{128} = 0.75,$ $\dfrac{72}{96} = 0.75,$ $\dfrac{54}{72} = 0.75,$ $\dfrac{40.5}{54} = 0.75,$ $\dfrac{30.375}{40.5} = 0.75.$

Each term is the previous term multiplied by 0.75.

Thus the sequence *is* a geometric sequence with first term = 128
the common ratio = 0.75
and T_n = $0.75 T_{n-1}$

c The given sequence is   4,   9,   14,   19,   24,   29,   …,

$\dfrac{9}{4} = 2.25,$ $\dfrac{14}{9} = 1.\overline{5}.$

The sequence does not have a common ratio.
Thus the sequence is *not* a geometric sequence.

EXAMPLE 5

$400 is invested in an account and earns $20 interest each year.

a How much is the account worth after 1 year, 2 years, 3 years and 4 years?

b Do the amounts the account is worth at the end of each year form an arithmetic sequence, a geometric sequence or neither of these?

Solution

a

Initial value	value after 1 year	value after 2 years	value after 3 years	value after 4 years	...
$400	$400 + 1($20)	$400 + 2($20)	$400 + 3($20)	$400 + 4($20)	...
$400	$420	$440	$460	$480	...

After 1, 2, 3 and 4 years the account is worth $420, $440, $460 and $480 respectively.

b The situation gives rise to amounts with a common difference of $20.

The amounts the account is worth at the end of each year form an arithmetic sequence.

As you are aware, the situation described in the previous example is not the way that an investment usually earns interest. Once the $20 interest has been added at the end of year 1 the account has $420 in it and it is this $420 that attracts interest in year 2, not just the initial $400. In this way the interest earned in one year itself attracts interest in subsequent years, i.e. compound interest is involved, rather than the simple interest situation described in the previous example.

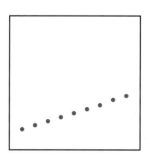
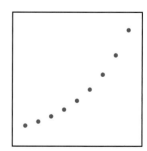

ISBN 9780170394994

EXAMPLE 6

$2000 is invested and accrues interest at a rate of 10% per annum, compounded annually.

a If no further deposits are made how much will be in the account after 1 year, 2 years, 3 years and 4 years?

b Do the amounts the account is worth at the end of each year form an arithmetic sequence, a geometric sequence or neither of these?

Solution

a

Initial value	value after 1 year	value after 2 years	value after 3 years	value after 4 years	...
$2000	$2000 × 1.1	$2000 × 1.1^2	$2000 × 1.1^3	$2000 × 1.1^4	...
$2000	$2200	$2420	$2662	$2928.20	...

After 1, 2, 3 and 4 years the account is worth $2200, $2420, $2662 and $2928.20 respectively.

b The situation gives rise to amounts with a common ratio of 1.1.

The amounts the account is worth at the end of each year form a geometric sequence.

Exercise 4A

For each of the following arithmetic sequences state

- the first term, T_1,

and • the $(n + 1)$th term, T_{n+1}, in terms of the nth term, T_n.

1	6,	10,	14,	18,	22,	26,	...
2	28,	26,	24,	22,	20,	18,	...
3	5,	15,	25,	35,	45,	55,	...
4	7.5,	10,	12.5,	15,	17.5,	20,	...
5	100,	89,	78,	67,	56,	45,	...

For each of the following geometric sequences state

- the first term, T_1,

and • the nth term, T_n, in terms of the $(n - 1)$th term, T_{n-1}.

6	6,	12,	24,	48,	96,	192,	...
7	0.375,	1.5,	6,	24,	96,	384,	...
8	384,	96,	24,	6,	1.5,	0.375,	...
9	50,	150,	450,	1350,	4050,	12 150,	...
10	1000,	1100,	1210,	1331,	1464.1,	1610.51,	...

For each of the following sequences , 11–22, state whether the sequence is arithmetic, geometric or neither of these two types.

11 2, 5, 11, 23, 47, 95, …

12 1, 5, 25, 125, 625, 3125, …

13 13, 14.5, 16, 17.5, 19, 20.5, …

14 50, 39, 28, 17, 6, −5, …

15 1, 1, 2, 3, 5, 8, …

16 128, 160, 200, 250, 312.5, 390.625, …

17 $T_{n+1} = 3T_n, T_1 = 3$.

18 $T_{n+1} = T_n + 6, T_1 = 2$.

19 $T_{n+1} = 3T_n + 5, T_1 = 1$.

20 $T_n = (T_{n-1})^2, T_1 = 7$.

21 $T_n = T_{n-1} - 8, T_1 = 2000$.

22 $T_n = (0.5)T_{n-1}, T_1 = 8$.

23 An AP has a first term of 8 and a common difference of 3. Determine the first four terms of the sequence and the recursive rule for T_{n+1} in terms of T_n.

24 An AP has a first term of 100 and a common difference of −3. Determine the first four terms of the sequence and the recursive rule for T_{n+1} in terms of T_n.

25 A GP has a first term of 11 and a common ratio of 2. Determine the first four terms of the sequence and the recursive rule for T_{n+1} in terms of T_n.

26 A GP has a first term of 2048 and a common ratio of 0.5. Determine the first four terms of the sequence and the recursive rule for T_{n+1} in terms of T_n.

27 The graph on the right shows the number of vehicles a company sold in a particular country each year from 2011 to 2014.

   a Verify that the figures for these years are in arithmetic progression.

   b With $N_{2011} = 6000$ write a recursive rule for N_{n+1} in terms of N_n.

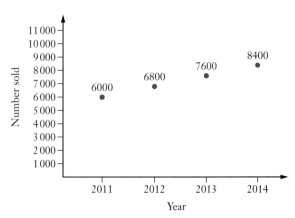

28 Each term of a sequence is obtained using the recursive rule

$$T_n = T_{n-1} + 10\% \text{ of } T_{n-1}$$

    a   Is the sequence an arithmetic progression, a geometric progression or neither of these?

    b   If the first term of the sequence is 500 find the next three terms.

29 Each term of a sequence is obtained using the recursive rule

$$T_n = T_{n-1} + 25\% \text{ of } T_{n-1}$$

    a   Is the sequence an arithmetic progression, a geometric progression or neither of these?

    b   If the first term of the sequence is 1000 find the next three terms.

30 Each term of a sequence is obtained using the recursive rule

$$T_n = T_{n-1} - 10\% \text{ of } T_{n-1}$$

    a   Is the sequence an arithmetic progression, a geometric progression or neither of these?

    b   If the first term of the sequence is 24 000 find the next three terms.

31 The graph on the right indicates the first five terms, T_1 to T_5, of a sequence, all of which are whole numbers.

    State   **a**   the first term and a recursive rule for the sequence,

            b   whether the sequence is arithmetic, geometric or neither of these types.

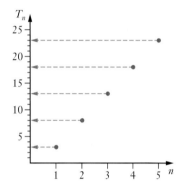

32 The graph on the right indicates the first five terms, T_1 to T_5, of a sequence. The last four of these terms are whole numbers.

    State   **a**   the first term and a recursive rule for the sequence,

            b   whether the sequence is arithmetic, geometric or neither of these types.

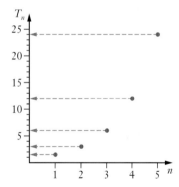

33 The graph on the right indicates the first five terms, T_1 to T_5, of a sequence. All of these terms are whole numbers.

    Determine   **a**   the first five terms,

    and   **b**   whether the sequence is arithmetic, geometric or neither of these types.

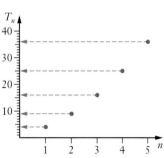

34 $1200 is invested in an account and earns $96 interest each year.

    a   How much is the account worth after 1 year, 2 years, 3 years and 4 years?

    b   Are these amounts in arithmetic progression, geometric progression or neither of these?

    c   Express the sequence of values:

                Initial value, value after 1 year, value after 2 years, …

    using recursive notation.

35 Won Yim starts working for a particular company on 1 January one year and is paid an initial annual salary of $45 000 with a guaranteed $1500 rise each year for the next 7 years.

Express the sequence of annual salaries over this time as a sequence using recursive notation and state whether the terms of the sequence progress arithmetically, geometrically or neither of these.

36 Joe started a new job on 1 January 2014 and, during 2014, he received a salary of $68 000. His contract guarantees a salary increase of 5% of the salary of the previous year on each subsequent 1 January, until and including 1 January 2017. Calculate Joe's salary for each year from 2014 to 2017.

Express the sequence of salaries from 2014 (term one) to 2017 (term four) using recursive notation.

37 $1500 is invested and accrues interest at a rate of 8% per annum, compounded annually. With this $1500 as the first term in the sequence express the value of the account on this and each subsequent year as a sequence defined recursively.

38 A taxi charges a fixed amount of $4.10 (called the flagfall) and then an additional $1.69 per kilometre. Determine the fare for a taxi journey of

    a   10 km,   **b**   20 km,   **c**   30 km.

39 (Flat rate, or straight line, method of depreciation.)

Over a period of 8 years a machine depreciates from its initial value of $138 000, to its scrap value of $26 000 at the end of the 8 years. If the amount depreciated each year stays the same determine the value of the machine after 1, 2 and 3 years.

40 (Unit cost, or units of production, method of depreciation.)

A particular photocopier is expected to have a lifetime of 3 million copies, after which its initial value of $27 000 will have reduced to a secondhand value of just $1500. If this depreciation is at a fixed amount per copy made, find the value of the photocopier after 500 000, 1 000 000 and 1 500 000 copies.

In its first year the photocopier makes 70 000 copies. What is its value at the end of this first year?

41 (Declining balance, or reducing balance, method of depreciation.)

Each year the value of a car depreciates by 15% of its value at the beginning of that year. The car is initially worth $36 000. With this $36 000 as the first term express the value of the car on this and each subsequent year as a sequence defined recursively.

Jumping to later terms of arithmetic and geometric sequences

Arithmetic sequences

Geometric sequences

Modelling arithmetic and geometric sequences

$$T_1, \quad T_2, \quad T_3, \quad \ldots \quad \ldots \quad T_{100}$$

Consider the arithmetic sequence defined by

$$T_{n+1} = T_n + 2 \text{ and } T_1 = 3$$

The rule allows us to obtain the terms of the sequence:

$$
\begin{aligned}
T_1 &= 3 \\
T_2 &= T_1 + 2 &= 3 + 2 &= 5 \\
T_3 &= T_2 + 2 &= 5 + 2 &= 7 \\
T_4 &= T_3 + 2 &= 7 + 2 &= 9 \qquad \text{etc.}
\end{aligned}
$$

However, if we wanted to know the value of a term much later in the sequence, say T_{100}, it would be a tedious process to have to calculate all of the terms up to T_{100}. It would be more useful if we could jump to the desired term without having to determine all of the preceding ones.

EXAMPLE 7

For the sequence defined recursively as $T_{n+1} = T_n + 7$ with $T_1 = 25$, determine the first four terms and the one hundredth term.

Solution

With $\quad T_{n+1} = T_n + 7 \quad$ it follows that

$$
\begin{aligned}
T_2 &= T_1 + 7 \\
&= 32. \\
T_3 &= T_2 + 7 \\
&= 39. \\
T_4 &= T_3 + 7 \\
&= 46.
\end{aligned}
$$

Note that for the second term we add 7 to T_1 *once*, for the third term we have added 7 to T_1 *twice*, for the fourth term we have added 7 to T_1 *three times*. It follows that for the one hundredth term we need to add 7 to T_1 ninety-nine times.

Hence
$$
\begin{aligned}
T_{100} &= T_1 + 99(7) \\
&= 25 + 693 \\
&= 718.
\end{aligned}
$$

The first four terms are 25, 32, 39 and 46 and the one hundredth term is 718.

If we apply the thinking used in the previous example to the general arithmetic sequence

$$a \qquad a+d, \qquad a+2d, \qquad a+3d, \qquad a+4d, \qquad a+5d, \qquad \ldots$$

we note that for T_2 the common difference, d, has been added once, for T_3 it has been added twice, for T_4 it has been added three times, and so on. Thus for T_n we need to add the common difference $(n-1)$ times.

Thus the arithmetic progression

$$a \qquad a+d, \qquad a+2d, \qquad a+3d, \qquad a+4d, \qquad a+5d, \qquad \ldots$$

has an nth term given by:

$$T_n = a + (n-1)d \qquad\qquad \text{Or } t_n = t_1 + (n-1)d$$

Just pause for a moment and check that you understand the difference between a recursive formula, which tells you how each term is obtained from the previous term, and the formula for the nth term, which allows you to jump to any term.

Note: Thinking of T_n as y, and n as x, the reader should see similarities between

$$T_n = a + (n-1)d \qquad \text{and} \qquad y = mx + c.$$

This is no surprise given the linear nature of arithmetic sequences.

EXAMPLE 8

For the AP: 11,   14,   17,   20,   ...

Determine   **a** T_{123},

b T_{500},

c   which term of the sequence is the first to exceed $1\,000\,000$.

Solution

a $\begin{aligned} T_{123} &= 11 + 122\,(3) \\ &= 377 \end{aligned}$

b $\begin{aligned} T_{500} &= 11 + 499\,(3) \\ &= 1508 \end{aligned}$

c   Suppose that T_n is the first term to exceed $1\,000\,000$.

Now $\qquad\qquad\quad T_n = 11 + (n-1)3$

$\therefore \qquad 11 + (n-1)3 \;>\; 1\,000\,000$

i.e. $\qquad\qquad\quad n \;>\; 333330.\overline{6}$

Thus the first term to exceed $1\,000\,000$ is $T_{333\,331}$.

EXAMPLE 9

An AP has a 50th term of 209 and a 61st term of 253. Find   **a**   the 62nd term,

   b   the 1st term.

Solution

a   To go from T_{50} to T_{61} we must add the common difference 11 times.

Thus if d is the common difference then

$$253 - 209 = 11d$$
$$\therefore \qquad 44 = 11d$$
$$\text{giving} \qquad d = 4$$
$$\text{Hence} \qquad T_{62} = T_{61} + 4$$
$$= 257$$

The 62nd term is 257.

b   From our understanding of APs it follows that

$$T_{50} = T_1 + 49d$$
$$\therefore \qquad 209 = T_1 + 49(4)$$
$$209 - 196 = T_1$$
$$T_1 = 13$$

The 1st term is 13.

Alternatively we could use the given information to write

$$a + 49d = 209$$
$$\text{and} \qquad a + 60d = 253$$

and then solve these equations simultaneously.

EXAMPLE 10

For the sequence defined recursively as $T_{n+1} = 1.5T_n$ with $T_1 = 8192$, determine the first four terms and the fifteenth term.

Solution

With $\qquad T_{n+1} = 1.5T_n \qquad$ it follows that

$$T_2 = 1.5T_1$$
$$= 8192 \times 1.5$$
$$= 12\,288$$
$$T_3 = 1.5T_2$$
$$= 12\,288 \times 1.5$$
$$= 18\,432$$
$$T_4 = 1.5T_3$$
$$= 18\,432 \times 1.5$$
$$= 27\,648$$

Note that for the second term we multiply T_1 by 1.5 once, for the third term we have multiplied T_1 by 1.5 twice, for the fourth term we have multiplied T_1 by 1.5 three times. It follows that for the fifteenth term we need to multiply T_1 by 1.5 fourteen times.

Hence $\qquad T_{15} = T_1 \times 1.5^{14}$
$$= 8192 \times 1.5^{14}$$
$$= 2\,391\,484.5$$

The first four terms are 8192, 12 288, 18 432 and 27 648.

The fifteenth term is 2 391 484.5.

Make sure you can obtain this same value for the fifteenth term of the sequence of the previous example by using a calculator to display terms of the sequence.

Are the larger numbers displayed as shown on the right or does your calculator use scientific notation to display these numbers?

Can your calculator display the terms of the sequence graphically?

Explore this possibility.

☑	$a_{n+1} = 1.5 \cdot a_n$
	$a_1 = 8192$
☐	$b_{n+1} = \square$
	$b_1 = 0$
☐	$c_{n+1} = \square$
	$c_1 = 0$

n	a_n
11	472392
12	708588
13	1062882
14	1594323
15	2391484.5

If we apply the thinking of the previous example to the general geometric sequence:

$$a, \quad ar, \quad ar^2, \quad ar^3, \quad ar^4, \quad ar^5, \quad ar^6, \quad \ldots$$

we note that T_2 is ar^1, T_3 is ar^2, T_4 is ar^3, etc. Hence $T_n = ar^{n-1}$.

Thus the geometric progression

$$a, \quad ar, \quad ar^2, \quad ar^3, \quad ar^4, \quad ar^5, \quad ar^6, \quad \ldots$$

has an nth term given by:

$$T_n = a \times r^{n-1}$$

Or $t_n = t_1 r^{n-1}$

EXAMPLE 11

Determine the 12th term and the 15th term of the geometric sequence:

$$0.0025, \quad 0.01, \quad 0.04, \quad 0.16, \quad \ldots$$

Solution

By inspection the common ratio is 4.

Hence the 12th term will be $\quad 0.0025 \times 4^{11} \quad = \quad 10\,485.76$
and the 15th term will be $\quad 0.0025 \times 4^{14} \quad = \quad 671\,088.64$

Again make sure that you can obtain these same answers using the ability of some calculators to display the terms of a sequence.

EXAMPLE 12

The 13th term of a GP is 12 288 and the 16th term is 98 304. Find **a** the 17th term,

b the 1st term.

Solution

a To go from the 13th term to the 16th term we must multiply by the common ratio 3 times.

If r is the common ratio then

$$T_{16} = T_{13} \times r^3$$

$\therefore$

$$98\,304 = 12\,288 \times r^3$$

Giving

$$r = 2$$

Hence

$$T_{17} = T_{16} \times 2$$
$$= 196\,608$$

The 17th term is 196 608.

b From our understanding of GPs it follows that

$$T_{13} = T_1 \times r^{12}$$

$\therefore$

$$12\,288 = T_1 \times 2^{12}$$

Giving

$$T_1 = 3$$

The 1st term is 3.

Alternatively we could write $ar^{12} = 12\,288$ and $ar^{15} = 98\,304$ and solve simultaneously.

Growth and decay

With their constant addition (or subtraction) arithmetic sequences describe linear growth (or decay).

With their repeated multiplication by r geometric sequences describe exponential growth, for $r > 1$, or decay, for $0 < r < 1$.

Consider the growth in the value of a house that is initially valued at $500 000 and is subject to an annual increase in value of 6.4%.

$$
\begin{array}{rcll}
\text{Initial value} & = & \$500\,000 & \leftarrow \quad T_1 \\
\text{Value after 1 year} & = & \$500\,000 \times 1.064 & \leftarrow \quad T_2 \\
\text{Value after 2 years} & = & \$500\,000 \times 1.064^2 & \leftarrow \quad T_3 \\
\text{Value after 3 years} & = & \$500\,000 \times 1.064^3 & \leftarrow \quad T_4, \quad \text{etc.}
\end{array}
$$

These values form a geometric sequence with

$$T_{n+1} = T_n \times 1.064$$

and

$$T_1 = 500\,000$$

Asked a question like: *At this rate how many years will it take for the value of this house to reach a value of $1 000 000?*

we could display the terms of our sequence on a calculator or spreadsheet and view the progressive year by year values, as shown on the next page.

	A	B	C	D
1	Initial value			$500,000.00
2	Percentage increase			6.40
3	Value at end of year		1	$532,000.00
4			2	$566,048.00
5			3	$602,275.07
6			4	$640,820.68
7			5	$681,833.20
8			6	$725,470.52
9			7	$771,900.64
10			8	$821,302.28
11			9	$873,865.63
12			10	$929,793.03
13			11	$989,299.78
14			12	$1,052,614.96

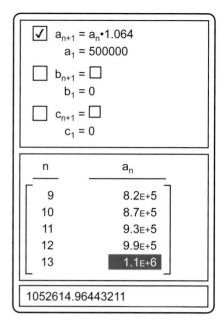

Create a spreadsheet like that shown above yourself.

The value of the house will be $1 000 000 shortly after the end of the eleventh year, i.e. early in the 12th year.

However note carefully that in this situation, with the recursive definition

$$T_{n+1} = T_n \times 1.064 \quad \text{and} \quad T_1 = 500\,000$$

T_1 is the value after zero years. Hence we must remember that if we use the ability of a calculator to generate the terms of the sequence, according to the recursive rule given above, then the balance after n years will be given by T_{n+1}. I.e. in the calculator display above, $n = 13$ gives the value at the end of 12 years.

One way to avoid this possible source of confusion would be to use the ability of some calculators to accept a sequence defined using T_0 as the 1st term.

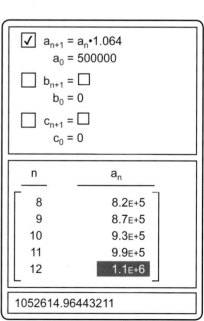

I.e. define the sequence as:

$$T_{n+1} = T_n \times 1.064 \quad \text{and} \quad T_0 = 500\,000,$$

as shown on the right.

Under such a definition T_n would indeed be the value after n years.

Alternatively we could use

$$T_{n+1} = T_n \times 1.064 \quad \text{and} \quad T_1 = 500\,000 \times 1.064$$

and again T_n would be the value after n years.

ISBN 9780170394994

Exercise 4B

Without using the sequence display routine available on some calculators determine the one hundredth term in each of the following arithmetic sequences.

1 11, 16, 21, 26, 31, 36, ...

2 −8, −5, −2, 1, 4, 7, ...

3 $T_{n+1} = T_n + 8$ with $T_1 = 23$.

4 $T_{n+1} = T_n - 2$ with $T_1 = 78$.

Without using the sequence display routine available on some calculators determine the twenty-fifth term in each of the following geometric sequences, leaving your answers in the form $a \times b^n$.

5 5, 10, 20, 40, 80, 160, ...

6 1.5, 6, 24, 96, 384, 1536, ...

7 $T_{n+1} = 3T_n$ with $T_1 = 8$.

8 $T_{n+1} = 2T_n$ with $T_1 = 11$.

Use the ability of some calculators to display the terms of a sequence to determine the requested term in each of the following sequences.

9 $T_{n+1} = T_n + 8$ with $T_1 = 7$.   Determine T_{28}.

10 $T_{n+1} = 35 - 2T_n$ with $T_1 = 5$.   Determine T_{20}.

11 $T_{n+1} = 3T_n + 2$ with $T_1 = 1$.   Determine T_{19}.

12 $T_{n+1} = (-1)^n T_n + 3$ with $T_1 = 6$.   Determine T_{45}.

13 Julie starts a new job at a factory manufacturing automobile components. The machine she operates requires several weeks before she is fully accustomed to it and so her output increases each day for the first 3 weeks (15 days). On the first day she successfully completes 48 items on the machine and increases this by 3 each day after that up to and including her 15th day on the machine.

Express the number of items completed on each of the first 15 days as a sequence using recursive notation.

How many items does she successfully complete on this 15th day on the machine?

14 Use the formula for the nth term of an AP with common difference d and $T_1 = a$, i.e. the formula $T_n = a + (n-1)d$, to explain why that for this AP, when we plot T_n on the y-axis and n on the x-axis the points obtained lie on a straight line of gradient d, and find the coordinates of the point where this straight line cuts the y-axis.

15 Use the formula for the nth term of a GP with common ratio r and $T_1 = a$ to explain why that for this GP, when we plot T_n on the y-axis and n on the x-axis the points obtained fit an exponential curve (i.e. a curve with an equation of the form $y = hk^x$ for constant h and k) and find the equation of this curve and the coordinates of the point where it cuts the y-axis.

16 Write a few sentences explaining what happens to the terms of the following arithmetic progression as n gets very large, (i.e. as n 'tends to infinity', which is written $n \to \infty$).

$$T_1 = a, \qquad T_2 = a + d, \qquad T_3 = a + 2d, \qquad T_4 = a + 3d, \qquad \ldots \qquad T_n = a + (n-1)d, \qquad \ldots$$

17 Write a few sentences explaining what happens to the terms of the following geometric progression as n gets very large, (i.e. as n 'tends to infinity', which is written $n \to \infty$).

$$T_1 = a, \qquad T_2 = ar, \qquad T_3 = ar^2, \qquad T_4 = ar^3, \qquad \ldots \qquad T_n = ar^{n-1}, \qquad \ldots$$

18 An arithmetic sequence has a first term of 8 and a common difference of 3. Determine the first four terms, the 50th term and the 100th term of the sequence.

19 An arithmetic sequence has a first term of 100 and a common difference of −3. Determine the first four terms, the 50th term and the 100th term of this sequence.

20 A geometric sequence has a first term of 11 and a common ratio of 2. Determine the first four terms, the 15th term and the 25th term of this sequence.

21 A geometric sequence has a first term of 2048 and a common ratio of 0.5. Determine the first four terms and the sixteenth term of this sequence.

22 Find an expression for T_n in terms of n for each of the following APs.
   a   9,   15,   21,   27,   33,   ...   **b**   7,   8.5,   10,   11.5,   13,   ...

23 Find an expression for T_n in terms of n for each of the following GPs.
   a   3,   6,   12,   24,   48,   ...   **b**   100,   110,   121,   133.1,   146.41,   ...

24 For the arithmetic sequence: 2,   9,   16,   23,   ...
   Determine   **a** T_{123},   **b** T_{500},
                    c   which term of the sequence is the first to exceed 1 000 000.

25 For the geometric sequence: 0.0026,   0.013,   0.065,   0.325,   ...
   Determine   **a** T_{12} ,
                    b   which term of the sequence is the first to exceed 1 000 000.

26 For the GP: 20 000 000,   15 000 000,   11 250 000,   8 437 500,   ...
   Determine   **a** T_{12} , giving your answer to the nearest hundred,
                    b   which term of the sequence is the first less than 1.

27 The nth term of a sequence is given by $T_n = n^3$.
   Obtain the first four terms of this sequence and state whether the sequence is arithmetic, geometric or neither of these.

28 An arithmetic sequence has a 19th term of 61 and a 41st term of 127.
   Find   **a**   the 20th term,   **b**   the 1st term.

29 An arithmetic sequence has a 50th term of 1853 and a 70th term of 1793.

Find   **a**   the 51st term,   **b**   the 1st term.

30 A geometric sequence has a 10th term of 98 415 and a 13th term of 2 657 205.

Find   **a**   the 14th term,   **b**   the 1st term.

31 A geometric sequence has a 7th term of 28 672, a 9th term of 458 752 and a negative common ratio.

Find   **a**   the 10th term,   **b**   the 1st term.

The following questions involve situations in which a quantity is repeatedly multiplied by an amount (geometric progression) or has the same amount repeatedly added to it, or subtracted from it (arithmetic progression). However, do not be too quick to reach for the formulae for the nth terms of such progressions. Instead think about what each situation requires you to find. This will often lead to the answer more quickly than simply 'clutching' at formulae.

32 $4000 is invested into an account paying interest at 8%, compounded annually. Determine (to the nearest cent) the amount in the account at the end of ten years.

33 a Scientists are studying a sample of bacteria and find that the initial population of 100 cells is doubling every hour.

How many cells will there be in this sample after 7 hours?

b At the end of the 7 hours the scientists introduce a chemical to the sample which not only immediately stops the growth of new cells but also kills off cells at a rate of approximately 400 per hour, for each of the next 5 hours. How many cells are present in this sample at the end of these 5 hours?

c Display the number of cells over this 12 hour period as a graph.

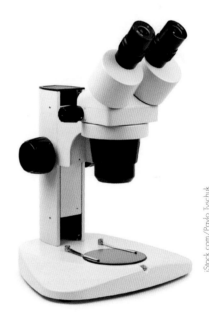

iStock.com/Pavlo Tyschuk

34 a The value of an item of machinery is expected to depreciate at a constant $5000 every year. Assuming this to be the case what will be the value of this item of machinery after 10 years if it is initially worth $98 000?

b By the end of each year the value of an item of machinery is expected to depreciate by 5% of its value at the beginning of that year. Assuming this to be the case what will be the value of this item of machinery after 10 years, to the nearest $1000, if at the start of this period it is worth $98 000?

35 The population of a country is growing exponentially. The populations in 2010, 2011, 2012 and 2013 were as follows.

Year	2010	2011	2012	2013
Population	45 million	45.8 million	46.6 million	47.5 million

Show that these figures support the theory that the population is growing at an annual rate of approximately 1.8%.

Predict the population for this country for the year 2027.

36 The population of a particular species of animal is declining exponentially. The numbers of these animals thought to be in existence in the wild in 2010, 2011, 2012 and 2013 were as follows:

Year	2010	2011	2012	2013
Population	18 000	16 500	15 200	14 000

Show that these figures support the theory that the number of these animals thought to be in existence is declining at an annual rate of approximately 8%.

Predict how many of these animals will exist in the wild in the year 2023.

37 When a particular ball is dropped onto a horizontal surface the height it reaches on its first bounce is 60% of the height from which it was dropped. Subsequent bounce heights are each 60% of the height of the previous bounce. If the ball is dropped from a height of 2 metres, onto a horizontal surface, find

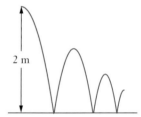

2 m

a the height reached on the first bounce,

b the height reached on the sixth bounce.

38 An investor purchases a valuable painting for $2 500 000. Experts predict that the value of the painting will rise exponentially at a rate of 4.6% per annum (i.e. the experts predict that by the end of each year the value of the painting will have risen by 4.6% of its value at the beginning of that year).

a Predict, to the nearest $10 000, the value of this painting after 5 years.

b Predict, to the nearest $10 000, the value of this painting after 10 years.

c Approximately how many years will it take for the value of the painting to reach $5 000 000?

39 If a house currently valued at $600 000 was to gain in value at 5.6% per annum, compounded annually, when would its value first exceed two million dollars?

ISBN 9780170394994

First-order linear recurrence relations

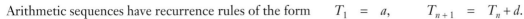
First-order recurrence relations

Arithmetic sequences have recurrence rules of the form $\quad T_1 = a, \qquad T_{n+1} = T_n + d.$

Geometric sequences have recurrence rules of the form $\quad T_1 = a, \qquad T_{n+1} = r \times T_n.$

We will now consider recurrence rules of the form $\qquad T_1 = a, \qquad T_{n+1} = b \times T_n + c.$

This last form is an example of a *first-order linear recurrence relation*.

- Arithmetic sequences and geometric sequences are also of the third form given above. With $b = 1$ and $c = d$ we have the AP rule and with $b = r$ and $c = 0$ we have the GP rule. However we will now consider more general examples of the third form.

- If the relationship involved the two previous terms, e.g. $T_{n+2} = b \times T_{n+1} + c \times T_n + d$ then it would be a *second-order linear recurrence relation*. We will concentrate our attention on first-order linear recurrence relations.

Consider recurrence relations for which each term, after the first, is obtained from the previous term by multiplying by some constant value, b, and then adding a constant value, c.

For example $\qquad T_1 = 400, \qquad T_{n+1} = 1.2 \times T_n - 50.$

In this case $\qquad T_1 = 400,$

$$
\begin{aligned}
T_2 &= 1.2 \times T_1 - 50 \\
&= 1.2 \times 400 - 50 \\
&= 430
\end{aligned}
$$

$$
\begin{aligned}
T_3 &= 1.2 \times T_2 - 50 \\
&= 1.2 \times 430 - 50 \\
&= 466
\end{aligned}
$$

$$
\begin{aligned}
T_4 &= 1.2 \times T_3 - 100 \\
&= 1.2 \times 466 - 50 \\
&= 509.2
\end{aligned}
$$

You might be wondering when such a first-order recurrence relationship would occur in real life.

Well let us consider the situation of a breeder of animals who, as well as increasing the size of his herd of these animals, also wants to sell some of his stock each year to other breeders.

Suppose the breeder has a herd of 400 animals and that through his successful breeding program this number is increasing at a rate of 20% per year. Suppose also that the breeder plans to sell 50 of the animals at the end of each year.

Initially his herd numbers $\qquad$ 400

One year later it will number: $\quad 400 \times 1.2 - 50 = 430$

After another year it will be $\quad 430 \times 1.2 - 50 = 466$

I.e. the recurrence relation considered previously (although the need to round to whole numbers would see the real situation deviate a little from the theoretical model).

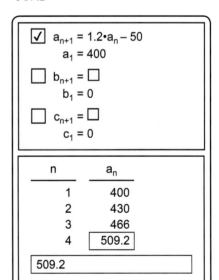

ISBN 9780170394994

Exercise 4C

1 Determine the first four terms for the sequence defined recursively as follows:
$$T_1 = 10, \qquad T_{n+1} = 2 \times T_n + 3.$$

2 Determine the first four terms for the sequence defined recursively as follows:
$$T_1 = 200, \qquad T_{n+1} = 3 \times T_n - 150.$$

3 Determine the first four terms for the sequence defined recursively as follows:
$$T_1 = 200, \qquad T_{n+1} = 2 \times T_n - 250.$$

4 For each of the sequences A, B and C defined below:

Sequence A: $\qquad T_1 = 5, \qquad\qquad T_{n+1} = 2 \times T_n + 1.$

Sequence B: $\qquad T_1 = 30, \qquad\qquad T_{n+1} = 2 \times T_n - 30.$

Sequence C: $\qquad T_1 = 90, \qquad\qquad T_{n+1} = 2 \times T_n - 95.$

a Determine the first five terms, T_1 to T_5.

b With T_n on the vertical axis and n on the horizontal axis, display T_1 to T_5 graphically.

c Choose the most appropriate label from the following list:

Increasing sequence   Decreasing sequence   Steady state sequence

5 Use your calculator to view the first 100 terms of the following recursively-defined sequence both numerically and graphically.
$$T_1 = 35\,000, \qquad T_{n+1} = 0.96 \times T_n + 2000.$$

What happens to the terms of the sequence 'in the long term', i.e. as $n \to \infty$?

Explain why this 'long term steady state situation' occurs.

6 A lakes complex is thinking of allowing some very limited fishing to occur in order to keep the population of fish in the lakes under control.

It is thought that the current population of fish is approximately 2000 and that if left uncontrolled this will increase at approximately 12% per year.

If limited fishing is allowed one suggestion is to allow approximately 300 fish to be caught near the end of each 12 month period with the first 'catch' occurring 12 months from now.

According to these figures, and if the limited fishing were allowed as suggested, what would the population of fish be just after the 'catch' in 1 year, 2 years, 5 years and 10 years?

7 A country has
- a current population of 17.6 million,
- a population growth rate (i.e. the birth rate take away the death rate) of approximately 1.2%,

and
- takes in roughly 120 000 people per year under its immigration program.

Writing the population of this country, n years from now, as T_n, write a recursive expression for T_n in terms of T_{n-1}, and state T_0.

Determine estimates for T_5 and T_{10}.

Miscellaneous exercise four

This miscellaneous exercise may include questions involving the work of this chapter, the work of any previous chapters, and the ideas mentioned in the Preliminary work section at the beginning of the book.

1 Write the next two entries in each of the following sequences.

a 3, 9, 27, 81, _____, _____.

b A, E, I, _____, _____.

c M, T, W, T, F, _____, _____.

d 32, 16, 8, 4, 2, 1, _____, _____.

2 Find the first six terms (T_1 to T_6) for each of the following.

a $T_{n+1} = 2T_n + 3$, $T_1 = 7$

b $T_{n+1} = 2T_n - 1$, $T_2 = 11$

c $T_{n+1} = 0.5T_n$, $T_3 = 64$

d $T_{n+1} = T_n + 7$, $T_4 = 26$

3 For the sequence 1, 3, 11, 43, 171, 683, 2731, 10923, ... find

a T_4 **b** $2T_3$ **c** $3T_1 + T_2$ **d** $T_3 - T_2$

e The sequence has a recursive formula of the form $T_n = aT_{n-1} + b$.

Find a and b and hence determine T_9 and T_{10}.

4 An arithmetic sequence has a first term of 10 and a common difference of 3.

Determine the 50th term of this sequence.

5 A geometric sequence has a first term of 1536 and a common ratio of 0.5.

Determine the 14th term of this sequence.

6 A study of a number of adults found that their heights, h cm, and their weights, w kg, could be reasonably well modelled by the equation

$$w = 0.9h - 80$$

Interpret the number 0.9 for this group of people.

7 Determine r_{xy}, the coefficient of linear correlation, and $y = A + Bx$, the equation of the linear regression line, for the following data.

x	49	26	29	40	34	22	35	37	40	29
y	54	41	44	50	44	42	51	48	55	47

Hence predict the value of y for   **a** $x = 32$,   **b** $x = 60$,

and comment on the validity of your predicted values.

8 With the aid of a calculator able to display the terms of recursively-defined sequences determine what happens to the terms of the following sequences 'in the long term', i.e. as $n \to \infty$.

a $T_1 = 5000$, $T_{n+1} = 0.8 \times T_n + 800$.

b $T_1 = 5000$, $T_{n+1} = 0.8 \times T_n + 1600$.

c $T_1 = 5000$, $T_{n+1} = 1.2 \times T_n - 800$.

9 $4000 is invested into an account paying interest at 8%, compounded annually and $200 is withdrawn from the account after each 12 months. Thus:

Amount in account at end of 1 year $= \$4000 \times 1.08 - \$200 \qquad \leftarrow \ T_1$

Amount in account at end of 2 years $= (\$4000 \times 1.08 - \$200) \times 1.08 - \$200 \quad \leftarrow \ T_2$

Express T_{n+1} in terms of T_n and, with the assistance of a calculator able to display the terms of a recursively-defined sequence, determine (nearest cent) the amount in the account at the end of ten years after the $200 for the year has been withdrawn.

10 The table below gives the number of motor vehicles and the number of televisions owned per 1000 people in eighteen countries.

Country	A	B	C	D	E	F	G	H	I
No. of motor vehicles (V)	714	450	203	361	570	92	748	202	528
No. of televisions (T)	731	618	412	542	486	215	815	271	560

Country	J	K	L	M	N	O	P	Q	R
No. of motor vehicles (V)	613	72	462	46	173	416	73	215	155
No. of televisions (T)	641	280	431	168	225	432	201	402	248

Display the above information on a scattergraph and use your graph to predict the number of televisions per 1000 people for a nineteenth country given that this nineteenth country has 285 motor vehicles per 1000 people.

Comment on the likely reliability of your prediction.

What percentage of the variation in the number of televisions can be explained by the variation in the number of motor vehicles?

11 A survey was conducted to investigate whether the frequency with which an adult engaged in sport or exercise was associated with gender. A number of people were asked how many times they engaged in sport or exercise in the course of an average week. The responses gave rise to the following table:

	Never	Once per week	Twice per week	Three times per week	Four times per week	Five times per week
Male	36	15	18	27	30	24
Female	52	18	31	33	46	40

Redraw the table showing either column percentages or row percentages as appropriate and determine, with justification, whether or not the results suggest that the frequency with which an adult engages in sport or exercise is associated with gender.

Shutterstock.com/Jacek Chabraszewski

ISBN 9780170394994

5.

Networks

Situation One: The bridges of Konigsberg

The city of Konigsberg, now known as Kaliningrad, features an island surrounded by the River Pregel. This island, and the other land masses involved, are connected by a number of bridges. In the 18th century the layout of the city was as shown in the diagram on the right with seven bridges connecting the various landmasses.

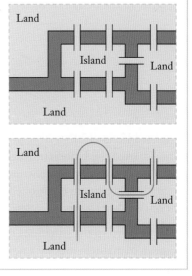

It is said that a popular activity of the time was to attempt to start at any of the landmasses and walk across all seven bridges, crossing each bridge once and once only.

One (failed) attempt is shown on the right.

Can it be done?

Situation Two

Is it possible to draw an unbroken curve to pass through all ten 'walls' in the diagram on the right without passing through any wall more than once?

Situation Three

The diagram below left shows the floor plan of a house.

The diagram below right shows one failed attempt to devise a route, starting and finishing at the same point, and that would pass through every door once and only once.

Is it possible? What if one more door could be added?

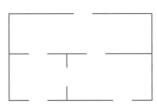

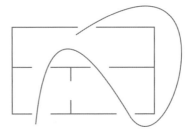

The three situations of the previous page may initially seem to have no relevance to real problems that may be faced in everyday life. However, finding the most appropriate route through a system is of great relevance in the design of such things as:

postal delivery routes,
rubbish collection routes,
house wiring systems,
computer networking systems,
routes for traffic wardens,
vehicle navigation systems, etc.

Leonhard Euler (1707–1783), in explaining whether the Konigsberg bridge problem was possible or not, started the branch of mathematics called **Topology**, of which the study of **networks** forms a part. This is not your first encounter with networks in this book because the road systems and social interactions that were mentioned in the *Preliminary work*, and that you would have met if you studied unit one of this *Mathematics Applications* course, were presented as network diagrams.

We will return to the situations of the previous page later in the chapter but first let us consider what we mean in mathematics by a network.

Networks

In mathematics a network is a set of points, called **vertices**, connected by a set of lines, called **edges**.

- The junctions and endpoints are called **vertices** (singular: **vertex**) or **nodes**.

4 vertices:

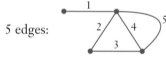

- The connections are called **edges**.

5 edges:

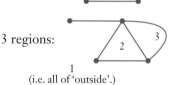

- The separated areas are called **faces** or **regions**. (Notice that we count the 'outside' region.)

3 regions:

(i.e. all of 'outside'.)

Check that you agree with each of the following:

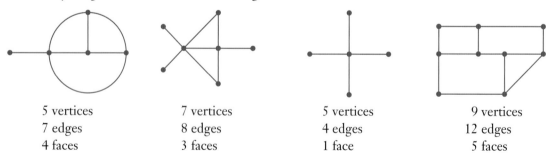

5 vertices	7 vertices	5 vertices	9 vertices
7 edges	8 edges	4 edges	12 edges
4 faces	3 faces	1 face	5 faces

Networks used in various situations are shown on the next page.

Chemical structure

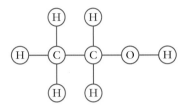

Social networks

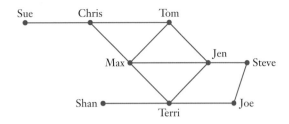

Technology network

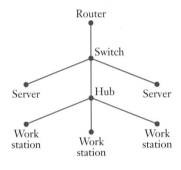

Rail system

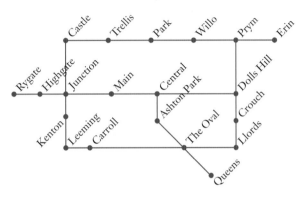

Exercise 5A

1 The diagram on the right is for a possible route that could be taken between Norseman and Perth.

Various locations on the route are shown with the distances between each of these as shown.

The diagram is not drawn to scale and the route is not necessarily straight.

For this route:

a How far is it from Kellerberrin to Merredin?

b How far is it from Northam to Coolgardie?

c If a steady speed of 100 km/hour could be maintained for the entire journey how long would the journey from Norseman to Perth take?

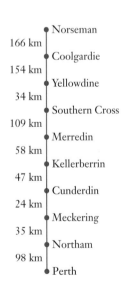

2 The diagram on the right shows the main connecting roads between six towns, A, B, C, D, E and F.

The diagram is not to scale but distances between each town, along the given roads, are shown in kilometres.

If we must always make progress towards the right, one possible route from A to F is ABCEF, for a total distance of 276 km. List all the other routes from A to F, and state the total distance for each route, remembering all travel must be towards the right.

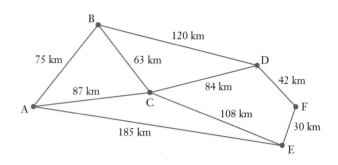

3 The diagram on the right shows the main buildings on a college campus, the possible cabled connections between them and the cost of each section of cabling.

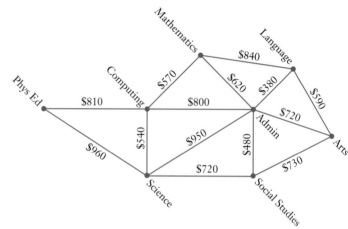

The second diagram shows one proposal for connecting each building to a central computer network by cabling.

This proposal has a total cost of $5140.

Try to come up with a cheaper system that still ensures that each building is 'on line'.

Can you find the cheapest system that still ensures that each building is 'on line'?

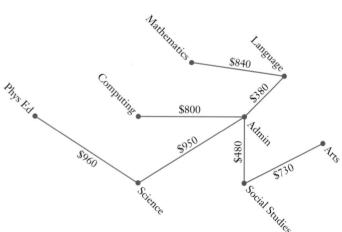

ISBN 9780170394994

How many vertices, edges and faces do each of the following networks have?

4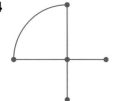

5

6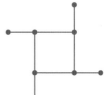

7

8

9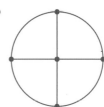

10

11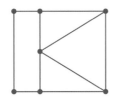

12

13

14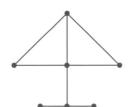

15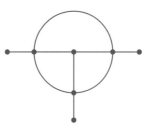

16

17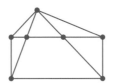

18

19

20

21

ISBN 9780170394994

There is a rule, called Euler's rule, connecting the number of vertices, edges and faces networks like those on the previous page have. Use your answers to questions 4 to 21 of the previous exercise to try to determine Euler's rule.

Traversability

We say that a network is **traversable** if it can be drawn without taking the pen off the paper and without going over the same edge twice (you may pass through the same vertex more than once).

The following networks *are* traversable. Check that you agree – this may not be as simple as it sounds because for some it is important where you decide to start.

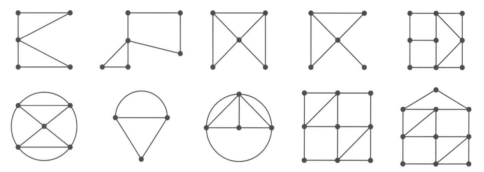

Some traversable networks can be traversed by starting and finishing at the same point.

Which of the traversable networks shown above have this property?

The following networks are *not* traversable. Again check that you agree.

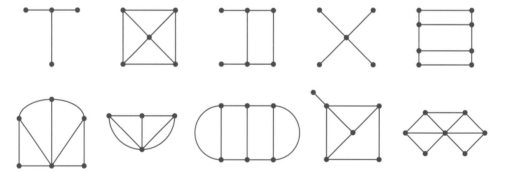

Can you discover a way of determining whether a network is traversable or not, without attempting to draw it?

Hint: Consider the number of edges meeting at each vertex.

The previous investigation of traversability may have reminded you of the activities at the beginning of this chapter. Let us consider those situations again.

The first of those situations involved the bridges of Konigsberg.

If we let the land masses that must be visited to be the vertices, shown as A, B, C and D in the diagram below left, and the bridges connecting them the edges, the situation can be represented as a network as shown below right.

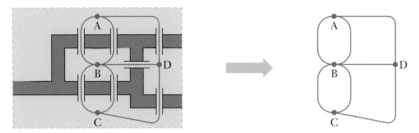

Is such a network traversable?

The second situation involved attempting to draw a curved line through all ten 'walls' in the diagram on the right.

We let the separate rooms, and the outside, be the vertices, as shown below left. The routes through a wall to another vertex can then be the edges. For example, if we are at A there are just two walls that if we pass through allow us to reach D without having to pass through other walls. Hence the diagram below right has two edges from A direct to D. Attempting to draw an unbroken curve through all ten walls, without passing through any wall more than once, then becomes the same as attempting to traverse the network.

Is such a network traversable?

In the third situation the challenge was to find a route that would pass through every door once and only once. Let the 'rooms' and 'outside' be the vertices and the doorways the edges that must be journeyed along:

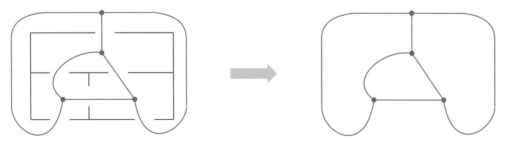

Is such a network traversable?

Graph theory

It is likely that for many students following this course a mention of the word *graph* makes you think of *x* and *y* axes, straight lines and curves, as shown below:

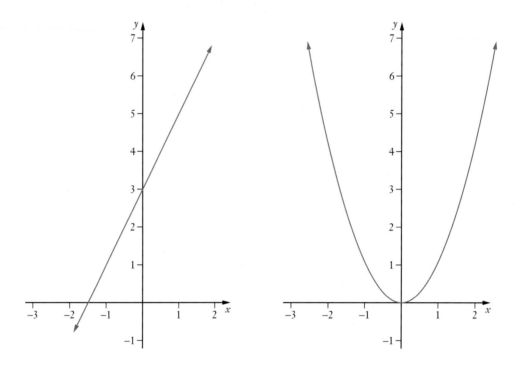

Perhaps some may also think of various statistical graphs:

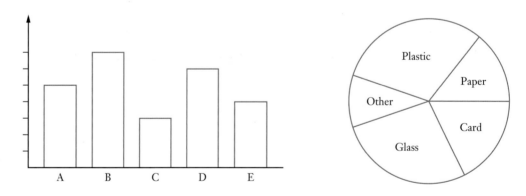

However, in this course we use the word graph in an even wider context to include structures used to display relationships and pairings. We therefore include network diagrams as **graphs**. Indeed the term **graph theory** tends to be taken to mean the study of networks.

ISBN 9780170394994

Some vocabulary

A graph is a diagram consisting of a set of points, called **vertices**, with some or all (or perhaps even none) of these vertices joined by a set of lines, called **edges**.

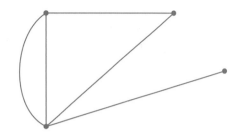

Any graph (or network) with an edge that starts and ends at the same vertex is said to contain a **loop**. In the diagram on the right, edge a_4 is a loop.

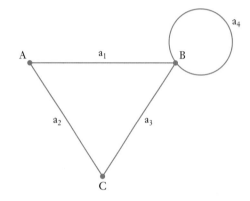

If two or more edges connect the same two vertices then the edges are said to be **multiple edges**.

Thus in the diagram on the right, a_2 and a_3 are multiple edges, as are a_4, a_5 and a_6.

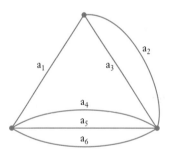

Some of the networks encountered earlier in this chapter had amounts or distances or some information on each edge, for example:

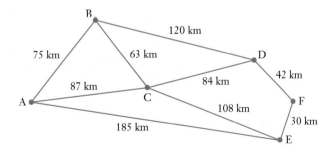

This is an example of a **weighted graph**, or **weighted network**. Each edge is worth a particular *weight*, be it distance, cost, time, etc.

As the *Preliminary work* section at the beginning of this text mentioned, if you have previously completed Unit One of *Mathematics Applications* you will be familiar with the idea of road systems and social interactions being displayed in the form of a network.

Suppose that the social interaction network, or graph, on the right shows which people, in a group of five people, A, B, C, D and E, have *shaken hands with* each other.

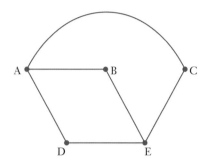

The network indicates that A has shaken hands with B, C and D but not E. C has shaken hands with A and E but not with B or D.

The action of shaking hands is directionless. If A has shaken hands with C then C has shaken hands with A.

Hence there are no directions shown on the network.

Suppose instead that the interaction network shown on the right shows who, in the group of five, has *been to the house of* someone else in the group. Now direction has significance. As the graph shows, E has been to the house of C but C has not been to the house of E. It makes sense for the graph to involve some **directed edges**, also referred to as **arcs**.

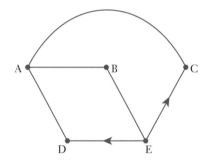

Similarly in the network of roads that exist linking towns A, B, C and D, as shown on the right, the arrows indicate that some roads are one way only.

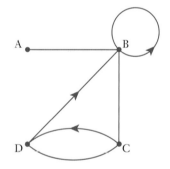

Again it makes sense for the graph to involve some directed edges.

Any graph having no directed edges is an **undirected**, or **non-directed**, graph.

Any graph involving directed edges is called a **directed graph**, also called a **digraph**.

ISBN 9780170394994

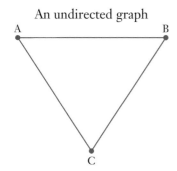

An undirected graph

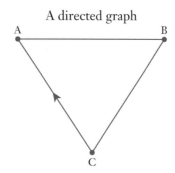

A directed graph

Note: In this unit, now that we are making the distinction between undirected graphs and directed graphs (digraphs), we will tend to show directions on all of the edges of a directed graph, rather than leave some 'unarrowed' as in the diagram above right. Hence in this book the digraph shown above right will tend to be drawn in one of the forms shown below.

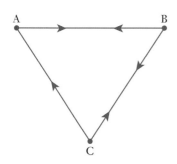

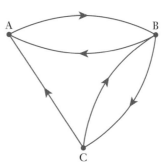

An undirected graph could be shown as 'directed' by placing opposing arrows on every edge, for example to show that all roads in a system are 'two way'. However such situations are more simply shown as undirected graphs.

> If we refer to a **simple graph** or **simple network** we mean an undirected and unweighted graph with no loops and no multiple edges.

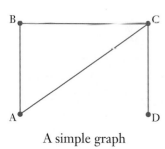

A simple graph

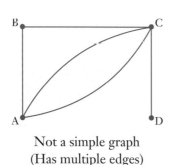

Not a simple graph
(Has multiple edges)

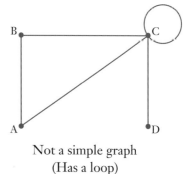

Not a simple graph
(Has a loop)

Note: Directed graphs and weighted graphs having no loops or multiple edges are sometimes referred to as **simple directed graphs** and **simple weighted graphs**. (Multiple edges in a digraph have the same start point and the same end point.)

Vocabulary continued – let's take a walk

Notice that in the graph, or network, shown on the right we could 'take a walk' from A to D via C. In graph theory, such a sequence of vertices, in this case A – C – D, in which there is an edge from each vertex to the next, is indeed called a **walk**.

In this network, ACD is not the only walk there is between A and D. Another would be ABCD.

Any walk that starts and finishes at the same vertex is called a **closed walk** whilst not ending at the starting vertex indicates an **open walk**.

For the network above, ABCD is an open walk whilst ABCA is a closed walk.

A walk may include repeated vertices and repeat use of edges, e.g. the walk ABCDCBC in the graph above, but if there is no repeat use of edges *and* no repeat use of vertices, other than perhaps ending at the same vertex as we start from, the walk is referred to as a **path**.

Thus for the network shown, ABCD is a **path** between A and D. In fact, to be more informative, ABCD is an **open path** as the starting vertex and the finishing vertex are different. In contrast, ABCA is an example of a **closed path** or **cycle**, because now the path finishes back at the starting vertex.

The '**length**' of a walk, path or cycle is the number of edges the walk, path or cycle uses. Thus the walk A – B – C – D – C – B – C has length 6 (repeat use of an edge counting each time), the path A – C – D has length 2 and the cycle A – B – C – A has length 3.

If a walk involves no repeated edges then we have a **trail**. In the network above, of the walks already mentioned, ACD, ABCD and ABCA are all trails but ABCDCBC is not due to the repeat use of some edges. However, while all paths are trails (because to be a path means no repeat use of edges *and* no repeat use of vertices whereas to be a trail only requires no repeat use of edges) we can have a trail that is not a path. For example, in the graph below, if we follow the edges in the sequence $a_1 a_2 a_3 a_4 a_5$ then ACBACD is a trail, but the repeat use of vertices A and C means that it is not a path.

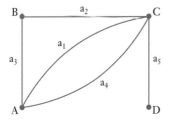

If a trail ends at the vertex it started from it is a **closed trail**.

Even more vocabulary – let's check the connections

In an undirected graph, if a path exists from one vertex, A, to another vertex, B, even if that path involves travelling along several edges, then the vertices A and B are said to be **connected vertices**. If the two vertices are connected by a path that does indeed only involve one edge then the vertices are said to be **adjacent vertices**. Hence adjacent vertices share a common edge.

Thus in the graph shown on the right:

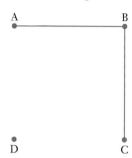

- A and B are connected vertices
  A and C are connected vertices
  B and C are connected vertices

- A and B are adjacent vertices
  B and C are adjacent vertices.

If, in an undirected graph (or network), every vertex is connected to every other vertex then the graph (or network) is said to be a **connected graph** (or **connected network**). Hence in the network above right, with there being no path from A to D (or from B to D or from C to D) the network is not a connected network. It is said to be **disconnected** or **unconnected**.

Each of the four simple networks shown below are *connected* networks:

 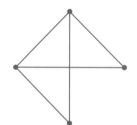 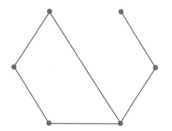 

Important note: The second graph of the four shown above is the first in this chapter that has shown edges crossing at a point that is not marked as a vertex. This is fine, simply imagine that one edge is passing over (or under) the other. If the edges were pipes or wires one pipe or wire passes above the other. In this way the graph is a two dimensional representation of a three dimensional situation. We will consider this aspect again later.

Each of the four simple networks shown below are *disconnected* networks:

 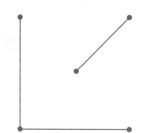 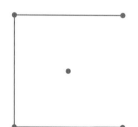 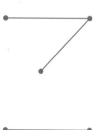

Note: With directed networks the situation is a little different and we would instead talk about the 'strength' of the connection. A search of the internet reveals a number of different conventions for describing connectedness in directed graphs. One system describes a directed network as being strongly connected, connected or weakly connected according to the following criteria:

> Strongly connected if for every pair of vertices there exists a path from one vertex to the other AND a path back.

> Connected if for every pair of vertices there is at least one path connecting the vertices, even if 'a way back' is not possible.

> Weakly connected if replacing all of the directed edges by undirected edges produces a connected undirected network.

Under this system, a strongly connected directed network is also connected and weakly connected. However we would tend to quote the strongest category and say it is strongly connected. Similarly a connected directed network is also weakly connected but we would simply say it is connected.

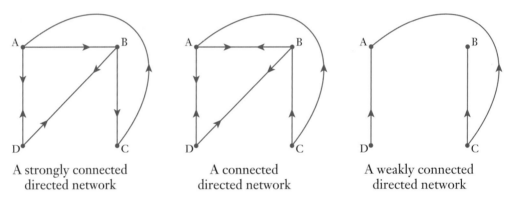

A strongly connected
directed network

A connected
directed network

A weakly connected
directed network

Notice that in the six simple connected networks shown below each vertex is connected to *every* other vertex in the network by a single edge. Or, to put it another way, we can make a journey from each vertex, along a single edge, to every other vertex. Or, to give a more technical statement, every pair of vertices are adjacent vertices.

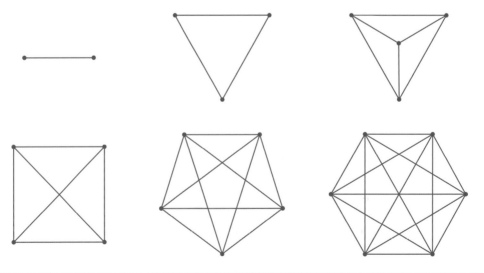

Simple graphs in which every vertex is connected to every other vertex by, in each case, a single edge, are called **complete** graphs.

Note: A complete graph with n vertices is sometimes denoted as K_n.

The previous page shows K_2 to K_6. (K_4 being shown twice.)

Consider the connected graph shown on the right.

Notice that we can remove any one of the seven edges in this graph and in each case the graph remains connected:

     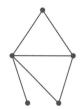 

Contrast this with the effect of the same single edge removal carried out on the connected graph that is now on the right.

(Note: The removal of an edge leaves all vertices in place.)

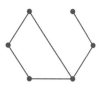

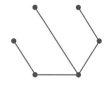  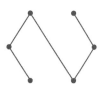

Notice that now, in some cases, removal of a single edge leaves the graph disconnected.

In a connected graph any edge which, when removed, leaves the graph disconnected, is called a **bridge**.

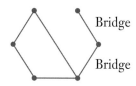

Bridge

Bridge

Summary!

Wow, the previous pages introduced a lot of new vocabulary! Whilst you should have managed to understand each word or expression as it was introduced you may now be having difficulty remembering exactly what each meant. Read through the following summary to revise the new vocabulary.

Graph theory involves the study of diagrams, called **networks** or **graphs**, consisting of a set of points, called **vertices**, with some or all (or perhaps even none) of these vertices joined by a set of lines, called **edges**. These edges can divide the graph up into a number of separate **faces** (or **regions**).

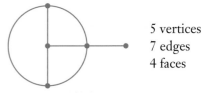

5 vertices
7 edges
4 faces

Any graph with an edge that starts and ends at the same vertex is said to contain a **loop**.

If two or more edges connect the same two vertices then the edges are said to be **multiple edges**. (If the two edges have direction then they are only multiple edges if they have the same direction between the same two vertices.)

If two vertices are connected by an edge then the vertices are said to be **adjacent vertices**.

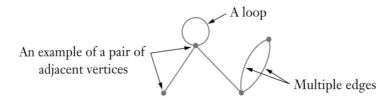

A loop

An example of a pair of adjacent vertices

Multiple edges

Graphs that have amounts or distances or some information on each edge are referred to as **weighted graphs**, or **weighted networks**.

Graphs that have **directed edges**, or **arcs**, are called **directed graphs** or **digraphs**.

Graphs with no directed edges are referred to as **undirected graphs**.

An undirected, unweighted graph with no loops and no multiple edges is called a **simple graph**. (Digraphs and weighted graphs with no loops and no multiple edges are sometimes referred to as **simple digraphs** and **simple weighted graphs**.)

A **walk** in a graph is a sequence of vertices for which each vertex in the sequence is joined to the next vertex in the sequence by an edge.

Any walk that starts and finishes at the same vertex is called a **closed walk** whilst not ending at the starting vertex indicates an **open walk**.

A walk may include repeated vertices and repeated edges.

If a walk involves no repeat use of edges and no repeat use of vertices, other than perhaps ending at the same vertex as we start from, the walk is referred to as a **path**.

A path that starts and finishes at different vertices is said to be an **open path**.

A path that starts and ends at the same vertex is said to be a **closed path** or **cycle**.

The 'length' of a walk is the number of edges it uses (repeat use of edges counted each time).

If a walk involves no repeated edges then we have a **trail**. However, while all paths are trails (because to be a path means no repeat use of edges *and* no repeat use of vertices whereas to be a trail only requires no repeat use of edges) we can have a trail that is not a path (if for example the trail involves a repeat vertex).

If a trail ends at the vertex it started with it is a **closed trail**.

Hence, for the graph shown on the right, the sequence of vertices AEC is not a walk (and therefore neither path nor trail) because there is no edge from A to E.

For this graph, whilst the following sequences of vertices are all walks

ABCD,   BCDEB,   CDECBA,

it would be more informative to say that

ABCD is an open path, of length 3.
(No repeated vertices, no repeated edges, hence a path.)

BCDEB is a closed path, or cycle, of length 4.
(No repeat vertices, apart from same start and finish, and no repeat edges.)

CDECBA is an open trail, of length 5.
(No repeat edges but a repeated vertex, hence a trail not a path.)

In an undirected graph, if a path exists from one vertex, A, to another vertex, B, even if that path involves travelling along several edges, then the vertices A and B are said to be **connected vertices**.

If, in an undirected graph, every vertex is connected to every other vertex then the graph is said to be a **connected graph**. If a graph is not connected it is **disconnected**.

In the graph shown on the right, paths exist from each vertex to every one of the other vertices. Hence every vertex is connected to every other vertex and so the graph is connected.

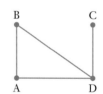

Any edge in a connected graph which, when removed, leaves a graph disconnected is called a **bridge**.

> With directed networks, or digraphs, the situation with regards to how the graph is connected is a little different. Instead we talk about the 'strength' of the connection as explained on an earlier page.

Complete graphs are simple graphs in which every vertex is connected to every other vertex by, in each case, a single edge. Three examples are shown below.

A complete graph with n vertices is sometimes denoted as K_n.

Exercise 5B

1 For each of the following state whether the graph is connected or disconnected.

(For **d** and **e** remember the 'important note' on page 119 regarding edges passing over, or under, each other.)

a

b

c

d 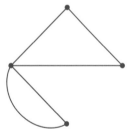 **e**

2 Defining a simple graph as an undirected, unweighted graph with no loops and no multiple edges, only two of the six graphs shown below are *simple* graphs.

State which two they are.

For each of those that are not simple graphs explain why not.

a

b

c

d  **e**  **f**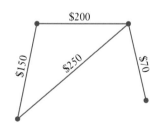

3 State the length of the paths and of the cycle shown below in heavy type.

a    **b**    **c**

4 For each of the situations below state which is likely to be the more appropriate form of representation – an undirected graph or a directed graph (a digraph) and explain your reasoning.

a A graph showing 'who likes whom' for a group of people.

b A graph showing 'who is aware of the existence of whom' for a group of people.

c A graph showing 'who is a Facebook friend of whom' for a group of people.

d A graph showing 'who is related to whom' for a group of people.

e A graph showing 'who has been to the movies with whom' for a group of people.

f A graph showing 'who lives within 5 km of whom' for a group of people.

g A graph showing 'who has the phone number of whom' for a group of people.

5 a Explain why the edge CE in the connected graph shown on the right is *not* a bridge.

b State which edges in the network shown are bridges and justify your answer.

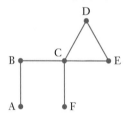

6 Which of the following simple graphs are complete graphs?

a   **b**   **c**

d    **e**   **f**

ISBN 9780170394994

7 We denote a complete graph involving *n* vertices as K_n.
Draw K_4 and K_8.

8 a For the network shown on the right, state which of the following sequences of vertices are walks.

AFE   ABCDE   BCD   ABDEF
ABDFA   ABDEFBC   BDEFB

b For each of the above sequences of vertices that are walks, describe each as one of

a path,   or   a cycle,   or   a trail,

whichever is the most informative in each case.

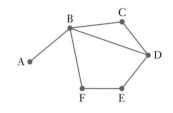

9 The graph on the right shows the relationship

is a proper divisor of

for the integers 1 to 8.

Construct similar graphs showing

is a proper divisor of

for   **a**   the integers 1 to 12,

         b   the integers 2 to 16.

10 Consider a tennis tournament involving three players in which each player plays a singles match against each of the other players (i.e. a 'round robin' tournament).

If we denote the players as A, B and C the tournament will involve the following three matches:

     A vs B,   A vs C,   B vs C.

This three match system can be represented by the complete graph K_3 shown on the right.

Let us suppose that the results of these matches were:

     A beat B   C beat A   C beat B

Adding directions to the previous graph allows these results to be included on the graph, as shown on the right.

Use this idea to show the following results from a five-player round robin as a digraph, with the direction of the arrow indicating who beat who.

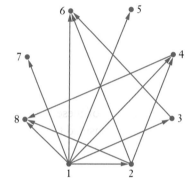

A beat B	C beat A	A lost to D	E lost to A	C beat B
B lost to D	B beat E	C beat D	C lost to E	D beat E

Planar graphs, Euler's rule, subgraphs, trees and bipartite graphs

Planar graphs

Let us consider again the three complete graphs shown below that featured on an earlier page.

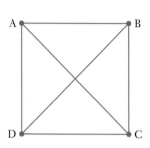
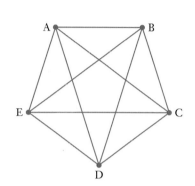
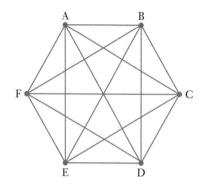

Notice that each of these graphs feature edges 'crossing over each other'. Where the edges cross are not vertices but instead they are places where one edge 'passes over' another. If these networks were, for example, systems of water pipes or wires in an electrical circuit, these locations would be where one pipe or wire passes over another. This 'one pipe above the other' situation means that the network has a place or places where the edges are not all 'in the same plane'.

We say that a graph that *can be* drawn in the plane, without its edges crossing over, is **planar**.

An important point to note here is that we say a network is planar if it *can be* drawn in the plane with its edges only intersecting at vertices.

The square-shaped network in the three shown above is planar because, whilst it hasn't been drawn without edges crossing over, it can be, as shown on the right.

Drawn in this way the network is more obviously planar.

As we saw earlier, this same planar network could be drawn as shown below.

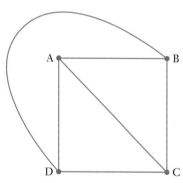

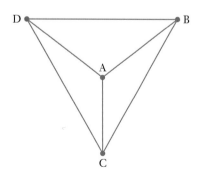

Consider again the network shown on the right.

In some situations it may be possible to insert a junction or vertex at the 'crossover point', as shown below:

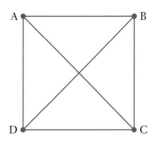

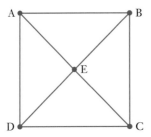

However, whilst the new system is more obviously *planar* it is no longer *complete* because there is now no single edge linking A and C (nor B and D). Also, in a real situation involving water pipes or electrical circuits, we may need AC and BD to be separate pipes or wires that do not form a junction with each other. Thus simply adding more vertices to give a planar system can change the original system significantly, especially in some real life contexts.

Euler's rule

Earlier in this chapter you were asked to determine Euler's rule using your answers for the number of vertices, edges and faces a number of graphs had. Did you manage to discover the rule?

The rule, for any connected planar network, is that if v is the number of vertices, f the number of faces and e the number of edges then:

$$v + f = e + 2$$
(I.e.: vertices + faces = edges + 2.)

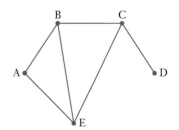

Subgraphs

Notice that each of the three graphs shown below involve a selection of the vertices and edges from the graph shown on the right.

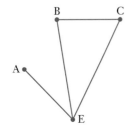

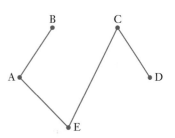

These three are all **subgraphs** of the original graph. (And indeed the middle graph of the three subgraphs is itself a subgraph of the one on the left.)

 ISBN 9780170394994

Trees

Any connected simple graph that contains no cycles is called a **tree**.

The following graphs are trees:

 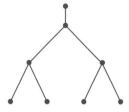 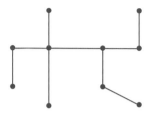

The following are not trees:

 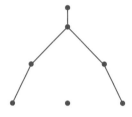 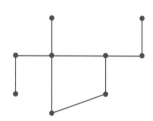

Contains a cycle
Therefore not a tree

Not connected
Therefore not a tree

Contains a cycle
Therefore not a tree

Note: A tree with n vertices will have $(n - 1)$ edges.
In a tree every edge is a bridge.

Bipartite graphs

Types of graphs

The graph below shows which of the three areas, midwifery (Mid), accident and emergency (A & E), and the intensive care unit (ICU), that three nurses are qualified to work in.

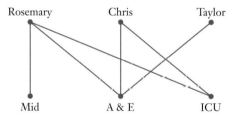

If we consider the vertices of this graph to be in two groups, Rosemary, Chris and Taylor being one group and Mid, A&E and ICU the other, each edge joins a vertex from one group to a vertex from the other group. There is no edge that by itself joins two vertices from the same group.

A graph in which the vertices can be split into two groups such that every edge joins a vertex from one group to a vertex of the other group, i.e. no edges join vertices from the same group, is called a **bipartite** graph. Thus, in a bipartite graph, every pair of adjacent vertices involves one vertex from one group and the other vertex from the other group.

(If a bipartite graph has every member of each group connected to every member of the other group it is a **complete bipartite graph**.)

ISBN 9780170394994

Exercise 5C

1   A connected planar graph has 5 vertices and 3 faces. How many edges does the graph have?

2   A connected planar graph has 4 faces and 6 edges. How many vertices does the graph have?

3   A connected planar graph has 5 vertices and 12 edges. How many faces does the graph have?

4   In this work on graph theory we are using the words vertices, edges and faces. However you have probably long been familiar with the use of these words when describing the features of some three dimensional shapes.

With the words used in this geometric, rather than graphical, sense determine the number of vertices, edges and faces each of the following three dimensional shapes have and then see if Euler's rule (Vertices + Faces = Edges + 2) is obeyed for each one.

a A cube	**b** A regular tetrahedron. (A polyhedron with four faces, each of which is an equilateral triangle.)	**c** A regular octahedron. (A polyhedron with eight faces, each of which is an equilateral triangle.)

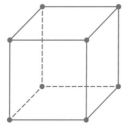
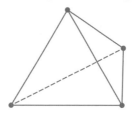
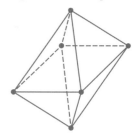

5   Which of the graphs shown below are subgraphs of the graph shown on the right?

For any that are not explain why.

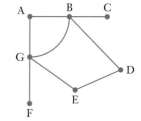

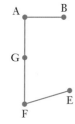
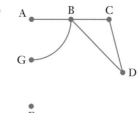
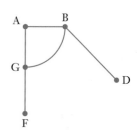

6 The table below shows which soccer clubs out of Manchester United, Manchester City, Liverpool, Everton and Arsenal, four soccer players, PB, BK, MO and FS played for (with a tick indicating the player did play for that club and a cross indicating the player did not).

	Arsenal	Everton	Liverpool	Manchester City	Manchester United
PB	✗	✓	✓	✓	✓
BK	✓	✓	✗	✓	✓
MO	✗	✗	✓	✗	✓
FS	✓	✗	✗	✗	✓

Show the information in the form of a bipartite graph.

7 For the graph shown on the right:

a Explain why the graph is not a complete graph.

b Show that the graph is planar by producing a version of it in which the edges only meet at vertices.

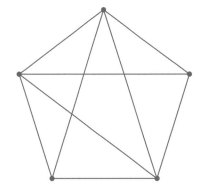

8 For the graph shown on the right:

a State whether the graph is a complete graph or not and explain your reasoning.

b Show that the graph is planar by producing a version of it in which the edges only meet at vertices.

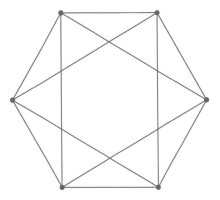

9 Show that the network drawn on the right is planar by producing a version of it in which the edges only meet at vertices.

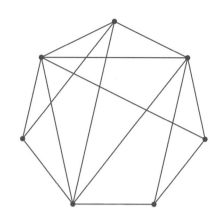

10 Given that the tree shown on the right is a bipartite graph what are the two vertex sets?

Draw the graph as a more obviously bipartite graph.

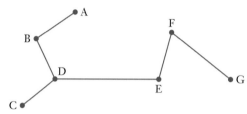

11 Is the graph shown on the right a bipartite graph? Justify your answer.

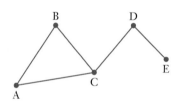

12 Is the graph shown on the right a bipartite graph? Justify your answer.

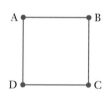

13 For each of the graphs shown below state whether the graph is bipartite.

If you claim that it is not bipartite justify your claim.

If you claim that it is bipartite redraw it as a more obviously bipartite graph.

a

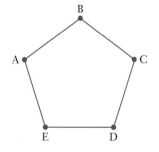

b

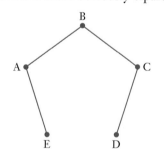

c

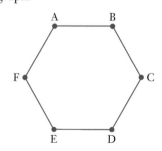

ISBN 9780170394994

14 a Explain why each of the 3 graphs shown below are 'complete' graphs.

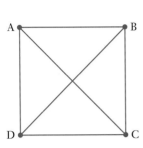 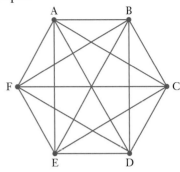

b For the first of the three complete graphs shown above it was explained some pages earlier that the graph could be redrawn with its edges only intersecting at vertices (i.e. without any edges 'crossing over') as shown on the right.

Can this be done for the other two complete graphs shown in part **a**?

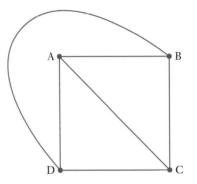

15 Suppose we have three houses, H1, H2 and H3 that the electricity company, the gas company and the water company want to connect up to their local connection points E, G and W.

H1   H2   H3
 •   •   •

 •   •   •
 E   G   W

Each company wants to give each house its own dedicated direct connection.

One way this can be done is as shown below.

However, the companies would prefer it if none of the edges crossed. Can this be done?
If not, can you draw a network with as few 'cross overs' as possible?

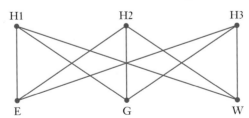

The degree or order of a vertex

In an undirected network we say that the degree or order of a vertex is the number of edges meeting at that vertex. Thus for the network shown on the right,

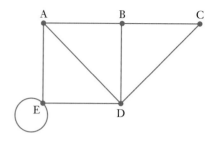

vertex A is of degree (or order)	3,
(sometimes written deg A = 3)	
vertex B is of degree (or order)	3,
vertex C is of degree (or order)	2,
vertex D is of degree (or order)	4,
and vertex E is of degree (or order)	4.

(Each end of the loop counts, hence the order of 4.)

We say that A and B are **odd vertices**, because for each of these vertices the order of the vertex is an odd number.

We say that C, D and E are **even vertices**, because for each of these vertices the order of the vertex is an even number.

You may well have considered such things when deciding if a network was traversable earlier in this chapter because:

For a network to be traversable it must be connected and have either zero odd vertices or exactly two odd vertices.

If a connected network has zero odd vertices a traversable route can start at any vertex and will finish at that same vertex.

If a connected network has exactly two odd vertices a traversable route must start at one of these and will then finish at the other.

Note: For *directed* graphs we have some edges coming into the vertex and other edges leaving the vertex. Thus we cannot simply talk of the order or degree of the vertices in a digraph. Instead we refer to an *in-degree* and an *out-degree*.

The 'in-degree' of a vertex is the number of edges 'coming in' to the vertex.

The 'out-degree' of a vertex is the number of edges 'going out' from the vertex.

Thus for the digraph shown to the right,

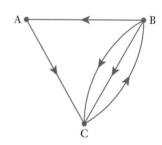

vertex A has an in-degree of   1,   and an out-degree of   1,
vertex B has an in-degree of   1,   and an out-degree of   3,
vertex C has an in-degree of   3,   and an out-degree of   1.

ISBN 9780170394994

Two particularly important network situations

A system of streets is as shown on the right. Suppose you are asked to organise the route that a street light inspection vehicle should take so that it starts and finishes at the vertex marked A and passes down each street once.

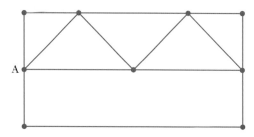

This is a traversability problem. It's like the Konigsberg bridge problem – travel along every edge once and once only, revisiting vertices if necessary, and ending back at the starting point.

The vertices in this graph are all even vertices so we should be able to find a suitable route.

A possible solution could be:

Start at A and do this street first.

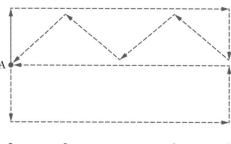

Contrast this with the situation in which the vertices represent all nine locations in a city that someone planning a walking tour wants to visit, starting and finishing at A.

Now we do not want to travel down all of the streets if we don't need to, we just want to visit every vertex.

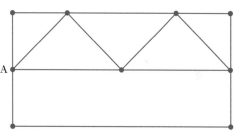

This is the sort of situation a travelling salesman might encounter with A representing his office and the other vertices representing locations he needs to visit. He wishes to visit every location and end up back at A, without visiting any location twice (except for his office which he will start from and end at). Now it is the vertices that must all be visited. If it can be done without travelling along particular edges that's fine.

Now a possible solution could be:

In the first situation, involving the street light inspection vehicle, we were looking for a closed trail that travelled every edge, i.e. one that finished at the same point it started from, that travelled every edge once and only once, vertices being visited more than once if necessary.

Eulerian graphs

A **trail** in a connected graph that travels every edge once and only once, repeated vertices permitted, is an **Eulerian trail**.

- If the trail is closed, i.e. starts and finishes at the same vertex, it is said to be an **Eulerian circuit** and the graph is said to be **Eulerian**. Hence a graph is Eulerian if it is traversable, starting and finishing at the same point.
  To be Eulerian a graph must have no odd vertices.

Hamiltonian paths and cycles

- If the trail is open, i.e. starts and finishes at different points, the graph is said to be **semi-Eulerian**. An open Eulerian trail is sometimes referred to as a **semi-Eulerian trail**. Hence a graph is semi-Eulerian if it is traversable, starting and finishing at different points. To be semi-Eulerian a graph must have exactly two odd vertices.

In the second situation, involving the walking tour, we were looking for a cycle that visited every vertex once and once only (except for the starting vertex that would be visited again upon completion). We did not have to travel every edge but did not want to travel any edge more than once.

A **path** in a graph that visits every vertex in the graph once only, with the possible exception of visiting the first vertex again at the end, thus making the path a cycle, is called a **Hamiltonian path**.

- If the Hamiltonian path is closed, i.e. begins and ends with the same vertex, the graph is said to be **Hamiltonian** and the closed path, or cycle, is said to be a **Hamiltonian cycle**.

- A connected graph that contains an open Hamiltonian path, but not a Hamiltonian cycle, is said to be **semi-Hamiltonian**.

Want to travel every edge exactly once, and don't mind repeating vertices?

Think Euler.

Want to visit every vertex exactly once (with the exception of maybe revisiting the starting vertex at the end), don't mind if some edges are missed, but don't want to travel any edges more than once?

Think Hamilton.

Leonhard Euler.	Swiss scientist and mathematician.	1707–1783.
William Hamilton.	Irish scientist and mathematician.	1805–1865.

Unfortunately, unlike the counting odd and even vertices that we can do when looking for Eulerian trails, there is no similar general test for the presence of Hamiltonian paths.

Check that you agree with the following:

Eulerian
Semi-Hamiltonian

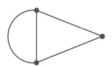

Semi-Eulerian
Hamiltonian

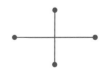

None of the
classifications

Exercise 5D

1 State the order of each of the vertices in the graphs shown below.

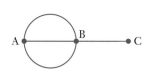 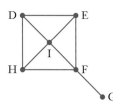 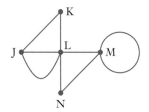

2 State the in-degree and the out-degree of each of the vertices shown in the following digraphs.

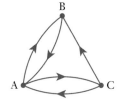 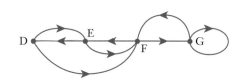 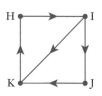

3 Classify each of the following graphs as Eulerian, semi-Eulerian or neither of these.

a   **b**   **c**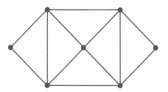

d 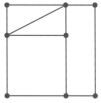 **e**  **f**

4 a Find an Eulerian circuit for the graph shown on the right, giving your answer as a sequence of vertices.

  b Find a Hamiltonian cycle for the graph shown on the right, giving your answer as a sequence of vertices.

5 a Is the graph shown on the right Eulerian, semi-Eulerian or neither of these?

  b Is the graph shown on the right Hamiltonian, semi-Hamiltonian or neither of these?

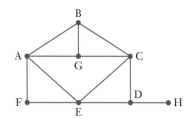

6 a Sketch an Eulerian circuit for the graph shown on the right, starting and finishing at vertex A.

   b Sketch an open Hamiltonian path for the graph shown on the right, starting at A and finishing at B.

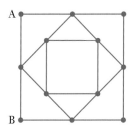

7 Find a Hamiltonian cycle for each of the following Hamiltonian graphs.

 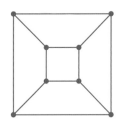 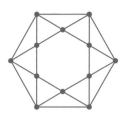 

Note: In some texts the definition of a semi-Hamiltonian graph leaves some doubt as to whether a graph can be both Hamiltonian and also semi-Hamiltonian. However, if we use the definition of a semi-Hamiltonian graph given on an earlier page in this book a graph cannot be both Hamiltonian and semi-Hamiltonian. The next two questions assume this latter situation to be the case.

8 Construct a bipartite graph like that shown below, where A, B, C, D and E refer to the graphs shown below the bipartite graph.

A   B   C   D   E

Eulerian Graph   Semi Eulerian Graph   Hamiltonian Graph   Semi Hamiltonian Graph

Graph A   Graph B   Graph C   Graph D   Graph E

ISBN 9780170394994

9 Construct a bipartite graph like that shown below, where V, W, X, Y and Z refer to the graphs shown below the bipartite graph.

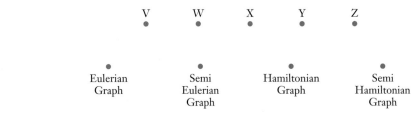

Graph V

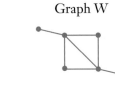
Graph W

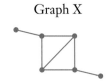
Graph X

Graph Y

Graph Z

10 Design a route around the road network shown below for a road inspection unit that needs to start at A, finish at A and pass down every road once and, if possible, only once.

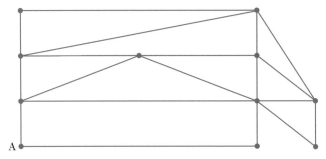

11 A contractor is given the job of re-marking the central road line on the system of roads shown below.

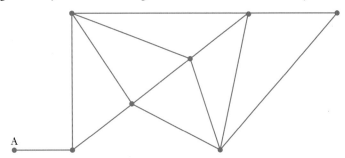

The contractor wants to design a route that starts at A, finishes at A and, if possible, passes along each and every road in the system once and once only. What route would you advise the contractor to take?

12 The organisers of a tourist brochure wish to show a circuit around the city that can be started at any of the eleven historically significant points in the city, takes in every one of the points and returns the walker to their starting point, having visited each of the points once. The graph below shows these eleven points as vertices and the pathways connecting them are as shown.

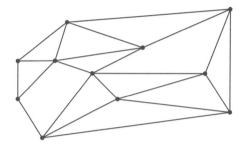

Design a suitable circuit.

Adjacency matrices

Adjacency matrices

As the *Preliminary work* section reminded us, the direct links in road networks and social interaction networks can be shown using a matrix form of presentation. For the digraph shown below left the associated matrix is as shown below right.

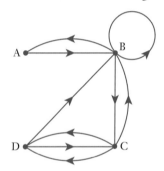

$$\begin{array}{c} \\ \text{From} \end{array} \begin{array}{c} \\ A \\ B \\ C \\ D \end{array} \overset{\displaystyle \text{To}}{\begin{array}{cccc} A & B & C & D \\ \left[\begin{array}{cccc} 0 & 1 & 0 & 0 \\ 1 & 1 & 1 & 0 \\ 0 & 1 & 0 & 2 \\ 0 & 1 & 1 & 0 \end{array}\right] \end{array}}$$

The entry in the second row and third column of the matrix, for example, shows the number of directed edges there are from vertex B to vertex C (in this case, 1).

Undirected graphs can similarly be presented in matrix form:

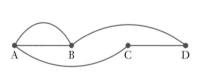

$$\begin{array}{c} \\ \text{From} \end{array} \begin{array}{c} \\ A \\ B \\ C \\ D \end{array} \overset{\displaystyle \text{To}}{\begin{array}{cccc} A & B & C & D \\ \left[\begin{array}{cccc} 0 & 2 & 1 & 0 \\ 2 & 0 & 0 & 1 \\ 1 & 0 & 0 & 1 \\ 0 & 1 & 1 & 0 \end{array}\right] \end{array}}$$

The two matrices shown above are examples of **adjacency matrices**. They inform us about the adjacency of pairs of vertices.

Notice that in the second case, with no loops present in the graph, the leading diagonal of the matrix shows only zeros, and with the graph being undirected the entries in the matrix are symmetrical across the leading diagonal.

For a graph with n vertices the adjacency matrix will be an $n \times n$ matrix.

In a directed graph
- the entry in row i, column j is the number of directed edges there are from vertex i to vertex j.

In an undirected graph
- the entry in row i, column j is the number of edges there are joining vertex i to vertex j,
- the adjacency matrix, A, will have $A_{ij} = A_{ji}$,
- a loop is counted as one edge. (See note below.)

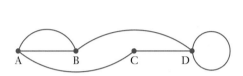

$$
\begin{array}{c}
\text{To} \\
\begin{array}{cc}
& \begin{array}{cccc} A & B & C & D \end{array} \\
\text{From} \quad \begin{array}{c} A \\ B \\ C \\ D \end{array} & \left[\begin{array}{cccc} 0 & 2 & 1 & 0 \\ 2 & 0 & 0 & 1 \\ 1 & 0 & 0 & 1 \\ 0 & 1 & 1 & 1 \end{array} \right]
\end{array}
\end{array}
$$

Note: In some applications a loop in an undirected graph may contribute '2', as we saw with the order of a vertex. However there we considered the loop's 'ends', rather like 'how many ends has a piece of string?'. Hence the contribution of '2'.

In the case of the adjacency matrix above we want the number of edges joining D to itself. There is 1 such edge. With edges undirectional it makes no sense to say the loop can be travelled 'one way or the other'. Hence the matrix shows 1, not 2.

However, whilst we will use this convention in this unit, as in the unit glossary of the syllabus at the time of writing, it is not necessarily universal. In some fields of graph theory the convention for undirected graphs is to count a loop as 2 in the matrix.

Consider the graph and corresponding adjacency matrix shown below:

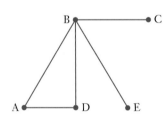

$$
\begin{array}{c}
\text{To} \\
\begin{array}{cc}
& \begin{array}{ccccc} A & B & C & D & E \end{array} \\
\text{From} \quad \begin{array}{c} A \\ B \\ C \\ D \\ E \end{array} & \left[\begin{array}{ccccc} 0 & 1 & 0 & 1 & 0 \\ 1 & 0 & 1 & 1 & 1 \\ 0 & 1 & 0 & 0 & 0 \\ 1 & 1 & 0 & 0 & 0 \\ 0 & 1 & 0 & 0 & 0 \end{array} \right]
\end{array}
\end{array}
$$

There is 1 walk of length 2 starting at A and finishing at C. The walk is: A B C.

There are 2 walks of length 2 starting and finishing at A. They are: A B A and A D A.

There are zero walks of length 2 starting at B and finishing at E.

You may remember from *Mathematics Applications Unit One*, these 'numbers of walks of length 2' can be obtained from the square of the above adjacency matrix.

Use your calculator to confirm that the matrix shown on the right is indeed the square of the adjacency matrix shown above. Check that you can see where the numbers of walks of length 2 mentioned above can be found in the matrix.

$$
\begin{array}{cc}
& \begin{array}{ccccc} A & B & C & D & E \end{array} \\
\begin{array}{c} A \\ B \\ C \\ D \\ E \end{array} & \left[\begin{array}{ccccc} 2 & 1 & 1 & 1 & 1 \\ 1 & 4 & 0 & 1 & 0 \\ 1 & 0 & 1 & 1 & 1 \\ 1 & 1 & 1 & 2 & 1 \\ 1 & 0 & 1 & 1 & 1 \end{array} \right]
\end{array}
$$

Exercise 5E

Write the adjacency matrix for each of the following undirected graphs.

1

A ————— B
| ╲ |
| ╲ |
D   C

2

A ————— B (with loop)
| ╲ |
D ————— C

3

A   B   C
 ╲ ╳ ╱
  ╳ ╳
 ╱ ╲
D   E

Write the adjacency matrix for each of the following digraphs.

4

A ——→ B
↑ |
| ↓
D ←—— C

5

A ——→ B
(double edges between A and D)
D   • C

6

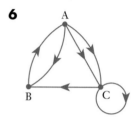

7 We denote a complete graph involving *n* vertices as K_n. (Remember, a complete graph is a simple graph in which every vertex is connected to every other vertex by, in each case, a single edge.)

Write adjacency matrices for K_3 and K_4.

8 The diagram on the right shows the direct flight links an airline operates between eight cities A, B, C, D, E, F, G and H.

Write the adjacency matrix for this system and confirm that your matrix is symmetrical across the leading diagonal.

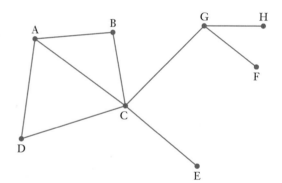

9 For the digraph shown on the right, and without using matrices, determine the number of walks of length 3 there are

   a   starting at A and finishing at C, listing each one as a sequence of vertices,

   b   starting at A and finishing at B, listing each one as a sequence of vertices,

   c   starting at B and finishing at A, listing each one as a sequence of vertices,

   d   starting at B and finishing at C, listing each one as a sequence of vertices.

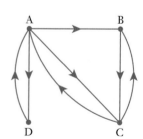

Write down the adjacency matrix, M, for the graph and, using your calculator to determine M^3, check your totals for parts **a** to **d**.

10 The network on the right shows the links between the cities A, B, C, D and E for which an airline operates direct flights. With a bit of thought you should be able to see from this graph that it is possible to fly with this airline between any two of these five cities in, at most, two flights.

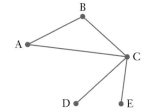

   a   Determine X, the adjacency matrix for this system.

   b   Determine $X + X^2$ and explain how this confirms the earlier statement that it is possible to fly with this airline between any two of these five cities in, at most, two flights.

11 A national communications network has ten 'hubs', A, B, C, D, E, F, G, H, I and J, through which various locations can be connected to other locations.

Not all possible pairs of hubs chosen from the ten have a direct connection between them.

In the adjacency matrix, X, shown below a '1' indicates a direct connection exists between two hubs, without the need for intermediate connections involving other hubs.

$$
X = \begin{array}{c} \\ A \\ B \\ C \\ D \\ E \\ F \\ G \\ H \\ I \\ J \end{array}
\begin{array}{c}
\begin{array}{cccccccccc} A & B & C & D & E & F & G & H & I & J \end{array} \\
\left[\begin{array}{cccccccccc}
0 & 1 & 0 & 0 & 0 & 1 & 0 & 0 & 0 & 0 \\
1 & 0 & 0 & 1 & 0 & 0 & 0 & 0 & 0 & 0 \\
0 & 0 & 0 & 1 & 0 & 1 & 0 & 0 & 0 & 0 \\
0 & 1 & 1 & 0 & 1 & 0 & 0 & 0 & 0 & 0 \\
0 & 0 & 0 & 1 & 0 & 1 & 1 & 1 & 1 & 0 \\
1 & 0 & 1 & 0 & 1 & 0 & 0 & 0 & 0 & 0 \\
0 & 0 & 0 & 0 & 1 & 0 & 0 & 0 & 0 & 1 \\
0 & 0 & 0 & 0 & 1 & 0 & 0 & 0 & 1 & 0 \\
0 & 0 & 0 & 0 & 1 & 0 & 0 & 1 & 0 & 0 \\
0 & 0 & 0 & 0 & 0 & 0 & 1 & 0 & 0 & 0
\end{array} \right]
\end{array}
$$

   a   How can we tell from the matrix that there are no direct connections between hubs that are 'one way only'?

   b   Using a calculator that can handle matrix manipulation, or an online matrix calculator, and not by drawing the graph, show that a connection can be made between every possible pairing with in each case, at most, four inter-hub connections, and that some connections between hubs cannot be done in less than four.

   c   Draw the graph and confirm that the situation described in part **b** is the case.

ISBN 9780170394994

Miscellaneous exercise five

This miscellaneous exercise may include questions involving the work of this chapter, the work of any previous chapters, and the ideas mentioned in the Preliminary work section at the beginning of the book.

1 For each of the networks shown below state whether the network is traversable or is not traversable.

Which of the networks are Eulerian?

Which are semi-Eulerian?

a

b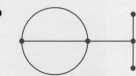

c

d

e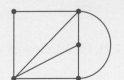

f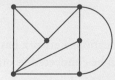

g

h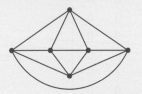

2 Is the sequence of account balances, in an account earning simple interest, an example of an arithmetic sequence, a geometric sequence or neither of these? Justify your answer.

3 Is the sequence of account balances, in an account earning compound interest, an example of an arithmetic sequence, a geometric sequence or neither of these? Justify your answer.

4 Find the first five terms of a sequence with a recursive formula:
$$T_n = 3T_{n-1} + 2, \qquad T_1 = 5.$$

5 Find the first five terms of a sequence with a recursive formula:
$$T_n = 2T_{n-1} + 3, \qquad T_1 = 5.$$

ISBN 9780170394994

6 With each room, the hallway and the outside area as vertices and the direct connections between them as edges, represent the following house layout as a network.

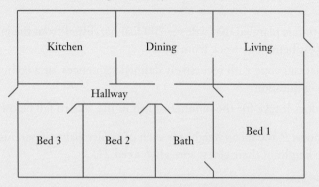

7 A survey determined that the animals frequenting a particular forest area were mainly insects, spiders, birds (including some hawks), snakes and some mongooses.

The food web showing 'who is being eaten by whom' for the area has been started below with the one arrow showing that some of the insects in the area are being eaten by some of the spiders in the area.

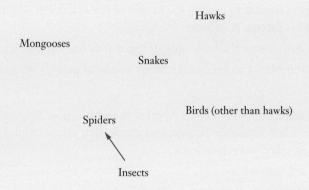

Copy the food web and complete it using your judgement of 'who is likely to be eaten by whom'.

8 In flat rate, or straight line, depreciation the value of an asset is depreciated by a fixed amount each year.

In reducing, or declining, balance depreciation the value of an asset is depreciated by a fixed percentage each year.

An asset that is initially worth $64\,000 has a depreciated value of $36\,000 two years later. What would be the value of this asset after one more year if the depreciation method used throughout is

a straight line depreciation?

b declining balance depreciation?

9 A commercial fish farming operation has approximately 10 000 fish in one of its artificial lakes. It is thought that without any intervention this population will continue to increase at the rate of 1.5% per week.

However, intervention is planned that will see 200 fish harvested from the lake at the end of each week, the first harvest being one week from now.

Use a calculator that can cope with recursively defined sequences, or a computer spreadsheet to answer the following question:

How many weeks does it take for the number of fish in the lake to fall below 8000?

10 The table below shows P, the mean height for each of twelve pairs of parents, all who have at least one son, and S, the height of their eldest son when aged 21.

P cm, mean parent height	168	179	175	179	161	183	170	172	174	186	175	163
S cm, eldest son height	170	186	172	185	162	191	181	175	179	199	176	171

a Determine the correlation coefficient r_{PS}, and the equation of the least squares regression line for predicting S from P.

b Use your equation from **a** to predict the height of the eldest son when aged 21 given that the mean height of his parents is 185 cm.

c Determine the equation of the least squares regression line for predicting P from S.

d Use your equation from **c** to predict the mean height of parents whose eldest son is 185 cm tall when 21.

11 Let us suppose that some mothers in two countries, A and B, whose eldest children were in the age range 12 to 16, were asked a number of questions about their eldest child. Further suppose that three of the questions asked were as follows:

Is your oldest child left handed?

Yes/No.

Would you say your child shows signs of being asthmatic? E.g. wheezing, dry irritating persistent cough, shortness of breath, asthma medication.

Yes/No

Has you child ever received treatment at a hospital for any reason, other than at birth?

Yes/No

The results were as shown in the table below.

	Question about left handedness		Question about asthma		Question about hospital treatment	
Country A	93 Yes	894 No	177 Yes	810 No	149 Yes	838 No
Country B	234 Yes	2179 No	651 Yes	1762 No	191 Yes	2222 No

Investigate, analyse, write a report.

ISBN 9780170394994

Shortest path

- Shortest path – a systematic approach
- Miscellaneous exercise six

Situation One

The diagram below shows a network of roads linking the locations, H, A, B, C, D, E, F and W shown as circles. (The circles are the vertices of the network, they are not loops. The reason for using open circles will become apparent later.) The diagram is not drawn to scale, and the roads are not necessarily straight, but the number on each road gives the road distance in kilometres between the locations (e.g. from H to A is 15 km).

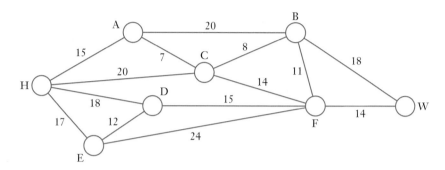

A motorist wishes to travel from home, at H, to work, at W, by the shortest route. Which route should the motorist take and how long will it be?

Situation Two

After travelling the shortest route a number of times the motorist of situation one realises that he wants to travel by the quickest route and, due to speed limits and road conditions, this may not be the shortest route. He uses his experience to estimate the time it takes him to travel each section of road in the network. These times, in minutes, are shown on the diagram below.

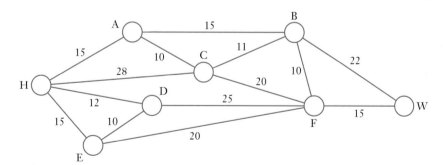

Assuming his estimates are correct which route should the motorist take to complete the journey from H to W as quickly as possible, and how many minutes will the journey take?

The situations on the previous page required us to find the **shortest path** from one vertex, point H, to another, point W. In situation one 'shortest' meant least distance and in situation two 'shortest' meant least time.

If the network is reasonably simple we may be able to identify the shortest path by inspection. For example, you should be able to identify the shortest path from A to D in each of the following by inspection.

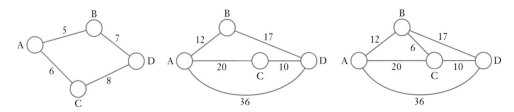

However, for a more complicated network this 'inspection' method may not be easy. For example, it would not be so easy to determine the shortest distance from A to Z in the network shown below simply by inspection. We may arrive at an answer but can we be sure it is the *shortest* path?

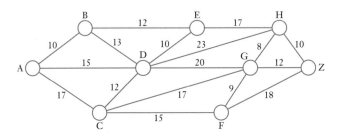

For such cases we need a process, or **algorithm**, to follow that will enable us to find the shortest path. (In mathematics an algorithm is a method of solving a problem by repeating a sequence of steps.)

Shortest path – a systematic approach

To determine the shortest path between two vertices, for networks in which this cannot easily be done by inspection, we proceed as shown in the example below.

Note: Previously we introduced the *length* of a path to mean the number of edges used, however, for weighted graphs it tends to be used to mean *total weight*.

EXAMPLE 1

Determine the length and the route of the shortest path from point A to point H in the network shown below.

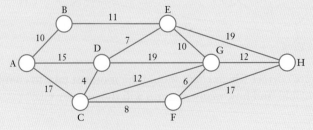

ISBN 9780170394994

Solution

The systematic approach we will consider involves moving from A, towards H, placing a number at each vertex in turn, to indicate the length of the shortest path from A to that vertex.

From A the shortest path to B is of length 10 units, to D is 15 units and to C is 17 units. Note in the diagram on the right the placement of these numbers and also note that the numbers on the edges giving these shortest routes are now indicated in red.

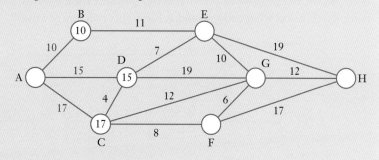

Note: While red printing is easy for a book to show, you are advised to circle the numbers on these edges using a red pen, or perhaps highlight the numbers with a highlighter pen.

If we continue this method to points E, F and G, the diagram will then be as shown on the right.

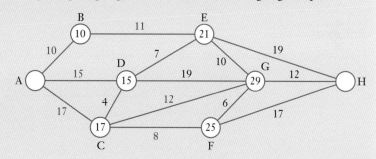

We can then complete the method by continuing on to point H:

Thus the shortest path from A to H is of length 40 units.

By backtracking from H back to A, along red numbers only we see that the shortest path has come from H–E–B–A.

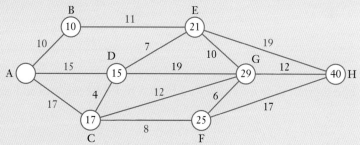

Thus the shortest path is 40 units and is achieved by the route ABEH:

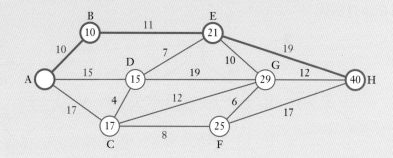

EXAMPLE 2

The network shown below shows the time, in minutes, taken to travel along various sections of road. According to these times what is the quickest time that the journey from A to I could be completed in and what route would achieve this quickest journey?

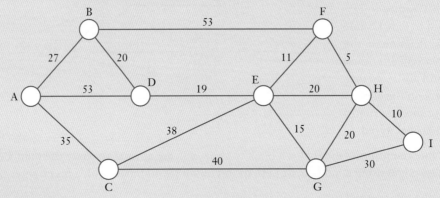

Solution

Following the systematic approach the network becomes 'marked up' as shown below.

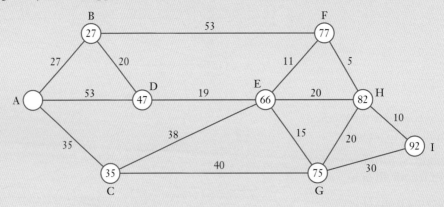

The shortest time for journeying from A to I is 92 minutes and is achieved by the route ABDEFHI.

Exercise 6A

For networks **1** to **3** shown below determine, by inspection, the length (i.e. total weight) and route of the shortest path from H to W.

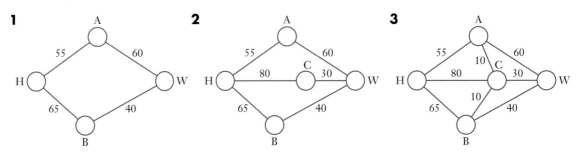

Use a systematic approach to determine the length (i.e. total weight) and the route of the shortest path from A to Z for the networks shown in questions **4** to **8**.

4

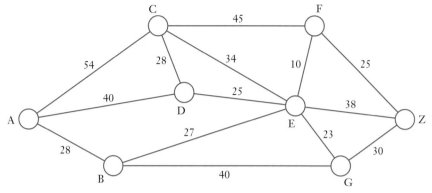

5

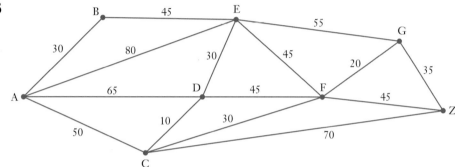

6

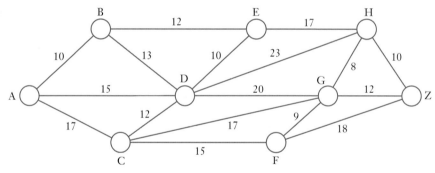

7

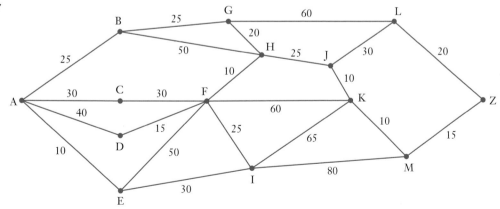

ISBN 9780170394994

8

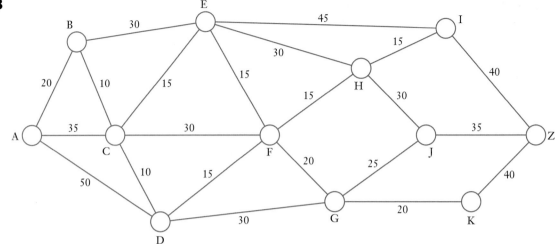

9 The network shown below shows the time, in minutes, taken to travel along various sections of road. According to these times what is the quickest time that the journey from A to K could be completed in and what route would achieve this quickest journey?

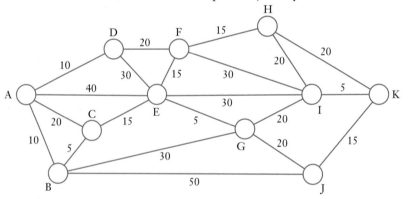

10 A decision is made to upgrade some of the roads between the villages of Ayesville and Hawton so that there is one completely upgraded route that motorists can take from Ayesville, all the way to Hawton. The diagram below shows the existing road sections and the numbers indicate the cost, in $10 000s, of upgrading each section. What route from Ayesville to Hawton should be chosen for upgrading if the cost of the upgrade is to be kept to a minimum? What will this minimum cost be?

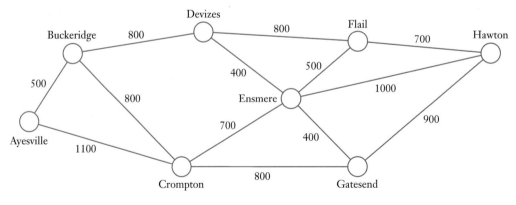

ISBN 9780170394994

11 The network shown on the right shows the road distances, in kilometres, between a number of places in Western Australia.

What is the shortest route, and how far is it, from

a Meckering to Wickepin,

b Meckering to Wagin?

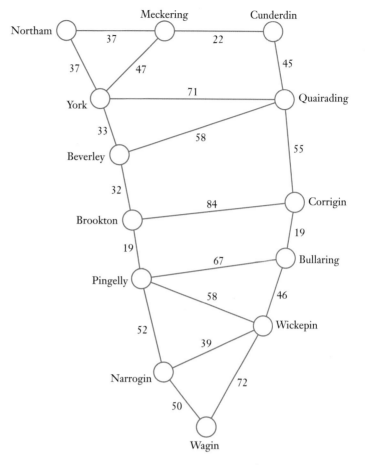

12 The network shown below represents the roads, shown as edges, linking a number of towns, shown as vertices. The numbers on each edge indicates the time, in minutes, required to travel that stretch of road.

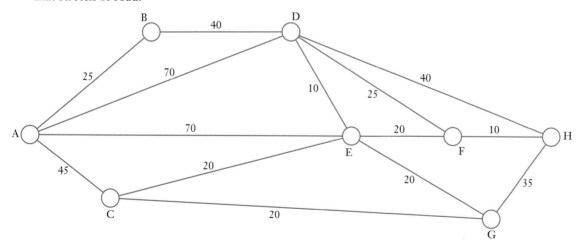

Determine the time and the route of the quickest journey possible from town A to town H, according to the given times.

13 The network shown below, not drawn to scale, shows the road distances, in kilometres, between a number of locations. What is the shortest route from Cataby to Morawa along the roads shown, and how far is this route?

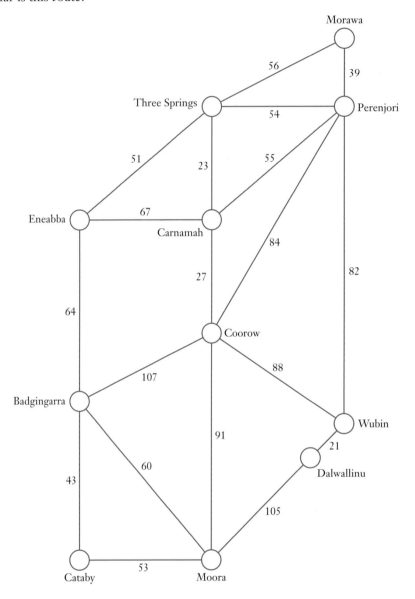

14 Find the shortest distance from A to I for the network on the right.

State the route that gives this shortest distance.

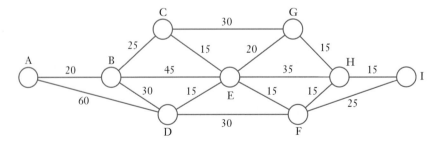

15 A hiker is planning her next day's journey. The network below shows the various tracks she can use to travel from her current position to the next campsite. Having an idea of the likely terrain and the condition of the tracks she estimates the time each section would take her, in minutes. According to these estimates (shown on the diagram) what route should she take in order to complete the journey as quickly as possible and, allowing an additional 30 minutes for a lunch break, how long should she expect the journey to take?

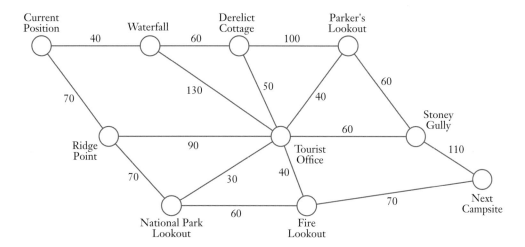

Having heard of the wonderful view at Parker's Lookout the hiker decides that she must include this in her route. What should her journey be now if it is to include Parker's Lookout and be completed as quickly as possible?

16 The diagram below shows the tracks running across a farming property. The owners of the property wish to have one route, from the main entrance to the house, upgraded. They ask a company to determine the cost of upgrading each track. Due to the variable quality of the existing tracks and the variable terrain that these tracks cross, the cost is not simply proportional to the length of the track. The numbers in the diagram indicate the cost of upgrading each track (in $1000). What route should they choose to minimise the cost and what would this minimum cost be?

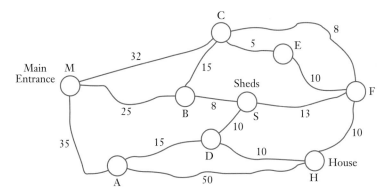

The owners then decide that the upgraded route must pass through junction S, where the farm sheds are situated. Does this requirement change the minimum cost and, if it does, determine the new route that incorporates the new requirement and keeps costs to a minimum? What would the minimum cost be now?

17 a Planners decide that a new road needs to be made through an area that is currently forestry. The planners decide that the best way to proceed is to upgrade some of the existing dirt tracks through the area to make one good route. The existing tracks are shown on the sketch of the area shown below (not drawn to scale) and the road is required from A to K.

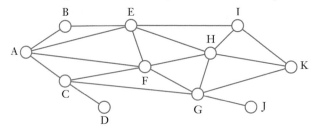

The length of each track, in metres, is given in the table below.

	A	B	C	D	E	F	G	H	I	J	K
A	–	210	320	–	600	510	–	–	–	–	–
B	210	–	–	–	420	–	–	–	–	–	–
C	320	–	–	140	–	280	460	–	–	–	–
D	–	–	140	–	–	–	–	–	–	–	–
E	600	420	–	–	–	180	–	210	350	–	–
F	510	–	280	–	180	–	240	200	–	–	–
G	–	–	460	–	–	240	–	240	–	350	440
H	–	–	–	–	210	200	240	–	180	–	500
I	–	–	–	–	350	–	–	180	–	–	350
J	–	–	–	–	–	–	350	–	–	–	–
K	–	–	–	–	–	–	440	500	350	–	–

What route from A to K should be chosen for upgrading to involve the minimum total length of track?

b Further investigation reveals that some of the existing tracks will cost more per metre to upgrade than others. Taking the track AF as the standard, the diagram below gives the 'cost per metre multiplication factor' for each track. For example to upgrade each metre of the track AC costs 1.3 times as much as it costs to upgrade 1 metre of AF.

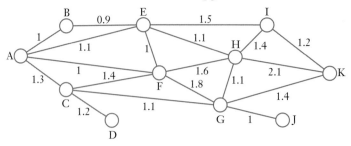

What route should be chosen for the upgrade in order to minimise cost?

c Before the upgrading commences a conservation group gets a ruling from the courts that the planned upgraded road must not pass through H. With this obeyed what route now gives the minimal cost?

ISBN 9780170394994

Miscellaneous exercise six

This miscellaneous exercise may include questions involving the work of this chapter, the work of any previous chapters, and the ideas mentioned in the Preliminary work section at the beginning of the book.

Write the adjacency matrix for each of the following undirected graphs.

1

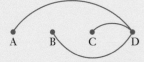

2

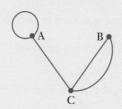

3

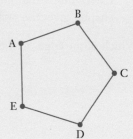

Write the adjacency matrix for each of the following digraphs.

4

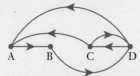

5

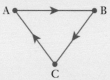

6

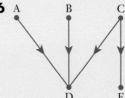

7 For question 6 above, simply by looking at the graph, determine what the square of the adjacency matrix will be.

8 The graph below was constructed for 5 flips of a fair coin.

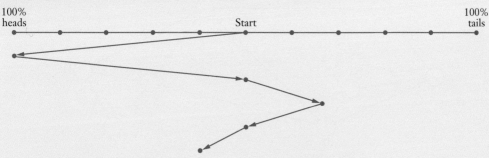

    a What was the sequence of heads and tails for these five flips?

    b Flip a coin 20 times and construct a similar diagram to that shown above for your results.

9 a Explain why it is correct to classify the graph on the right as being semi-Eulerian.

    b What edge needs to be added to the graph to make it Eulerian? Justify your answer.

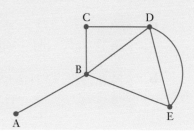

10 a For the weighted network shown on the right list all possible paths that start at A and finish at E, and state the total weight of each path. (Remember a *path* from A to E will have no repeat visits to vertices and no repeated travel along an edge.)

b Which path from A to E has the least total weight?

c Check that using the shortest path algorithm of this chapter gives the same answer as that obtained in part **b**.

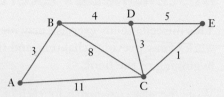

11 Which of the edges in the graph shown below are bridges?

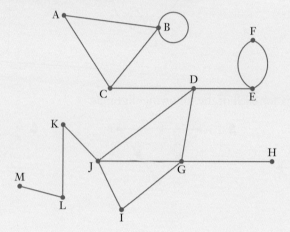

12 Find the length (i.e. total weight) and route of the shortest path from A to H in the following weighted graph.

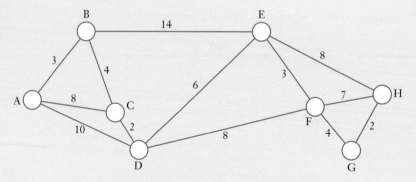

13 In each of the following sets of paired values what percentage of the variation in B can be accounted for, or explained by, the changes in A?

a

A	17	23	41	65	68	94	101	110	131	131	142	155
B	105	108	90	94	63	70	32	54	63	48	31	25

b

A	0.8	1.1	1.9	2.3	3.5	4.7	4.8	6.2	7.1	8.7	9.5	7.2
B	4.2	7.3	13.8	5.6	13.6	4.8	17.3	9.3	14.8	13.2	18.2	6.1

ISBN 9780170394994

14 The number of stitches in each row of a particular crochet pattern are as shown on the right.

1st row	⊕	⊕	⊕	⊕		
2nd row	⊕	⊕	⊕	⊕	⊕	
3rd row	⊕	⊕	⊕	⊕	⊕	⊕
4th row	⊕	⊕	⊕	⊕	⊕	⊕	⊕

With T_1 the number of stitches in the first row, T_2 the number in the second row, etc., express the sequence

$$T_1, \quad T_2, \quad T_3, \quad T_4, \quad T_5, \quad \ldots$$

using recursive notation (i.e. state T_1 and the recursive rule).

15 Determine by how much the eleventh term exceeds the tenth term in each of the following recursively defined sequences.

a $T_1 = 15,$ $\qquad T_{n+1} = T_n + 7.$

b $T_1 = 534,$ $\qquad T_{n+1} = T_n + 17.$

c $T_1 = 5,$ $\qquad T_{n+1} = 2 \times T_n.$

d $T_1 = 3,$ $\qquad T_{n+1} = 2 \times T_n + 4.$

16 Which of the following could be drawn as more obviously bipartite graphs?

a 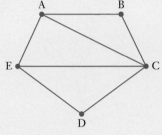 **b** 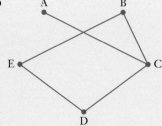 **c**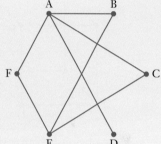

17 Fourteen students sat a mid-year exam and an end of year exam in a particular course. The results they gained were as follows:

Initials	KA	CB	JE	TE	TH	AJ	TK	RL	KM	NN	TP	DS	JS	RS	Out of
Mid Yr (M)	52	41	52	76	84	93	44	63	81	48	83	74	63	69	100
End Yr (E)	84	61	73	72	95	107	66	74	96	59	102	78	82	88	120

A fifteenth student missed the mid year exam but scored 65 out of 120 in the end of year exam.

a Determine the correlation coefficient for the 14 pairs of results given above.

b Determine the equation of the least squares regression line,

$$M = aE + b,$$

for predicting mid year marks based on end of year marks.

c Use your regression line from **b** to estimate a mid year exam mark for the fifteenth student.

18 The diagram below shows the existing paths, tracks and roads through part of a forest. Some of these will be upgraded to produce one good quality road from point A to point Z. The numbers on the edges in the diagram show the cost, in thousands of dollars, of upgrading each path, track or road to the standard required. Determine the cost and route of the cheapest upgrade that will give one good quality road from A to Z.

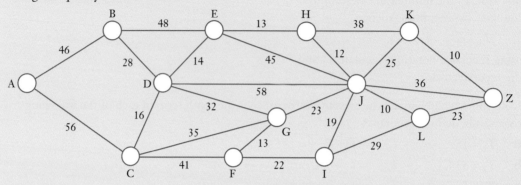

19 The graphs below show a cube, a tetrahedron and an octahedron.

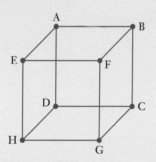 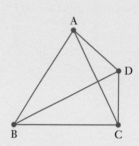 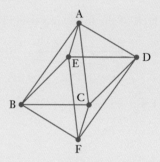

More obviously planar versions of these graphs are shown below.

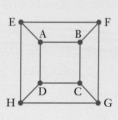 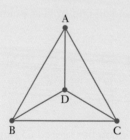 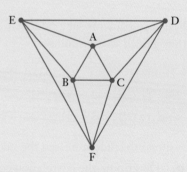

Of these three planar graphs which are

a connected graphs?

b complete graphs?

c Eulerian graphs?

b Hamiltonian graphs?

20 The table below shows the times in minutes that it takes to travel between eight towns along the direct road links that exist between any two of the towns.

	Alexville	Barton	Chelmes	Dane	Erinville	Felpam	Grange	Heslop
Alexville	–	20	25	45	–	–	–	–
Barton	20	–	–	20	–	60	–	–
Chelmes	25	–	–	30	40	–	–	–
Dane	45	20	30	–	45	35	40	–
Erinville	–	–	40	45	–	–	20	50
Felpam	–	60	–	35	–	–	10	–
Grange	–	–	–	40	20	10	–	20
Heslop	–	–	–	–	50	–	20	–

The 'map' on the right gives a rough idea of where the towns are situated with respect to each other.

Determine the shortest time that the journey from Alexville to Heslop can be completed in and state the towns you would pass through if you made the journey along the route that gives this shortest time.

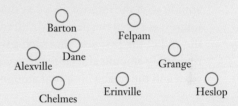

21 The 437 students who completed a particular business management course came from two different colleges, Stack and Bosun.

Every one of the 437 students received a grade of either A, B, C, D or E.

241 of the students were from Stack and 77 of these received an A grade.

Altogether 145 B grades were awarded which was 44 more than the total number of A grades awarded.

The number of Stack College students receiving a B grade was 16 more than the number from Stack receiving an A.

Of the 158 students receiving a C grade, 60 were from Stack.

9 of the Stack students received a D.

The number of Bosun students receiving an E was 4 times the number of Stack students receiving an E.

Copy and complete the following two-way table showing the number of students from each college achieving each grade.

	A	B	C	D	E	Totals
Stack College						
Bosun College						
Totals						437

Investigate whether there seems to be an association between the grade achieved and which of the two colleges the student comes from.

The data given in this question is fictional and does not relate to any colleges in real life.

22 The weighted graph shown on the right gives the estimated times, in minutes, required to travel on foot between the various pairs of locations that can be chosen from four locations, A, B, C and D situated in a challenging, rugged, outback environment. One possible Hamiltonian circuit starting and finishing at A is ACBDA with a total estimated time of 259 minutes.

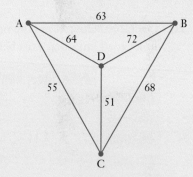

a For the estimated times given is this the quickest Hamiltonian circuit for this graph? Investigate.

b In this situation, but not necessarily with the times as stated, if we must start at A, visit each of B, C and D and then end up back at A, will it always be a Hamiltonian circuit that gives us the shortest time or will it depend on the estimated times?

23 A person opens an account with a deposit of $400 000. The account pays interest of 6.7% per annum with the interest added at the end of each twelve month period. At the same time as interest is added the person withdraws $55 000 per year for as many years as possible and, at the end of the first year that such a withdrawal is not possible, the person plans to withdraw the remaining amount and close the account. For how many years can the $55 000 be withdrawn and what will the final withdrawal amount be?

Extension

24 A farmer is developing his herd of a particular species of animal that is not commonly farmed, with the intention of selling them for meat and other products. The farmer currently has 500 animals in the herd and, allowing for normal birth and death rates, anticipates an annual growth of 10% (compound) before any are culled for market.

The farmer wishes to consider the following ideas.

- Cull a regular number at the end of each year, say 40.

- Cull an increasing number each year, say

40	at the end of year 1,
80	at the end of year 2,
120	at the end of year 3,
160	at the end of year 4, etc.

- Cull an increasing number each year, say

40	at the end of year 1,
40×1.2	at the end of year 2,
40×1.2^2	at the end of year 3,
40×1.2^3	at the end of year 4, etc.

Explore each of these possibilities in terms of their effects on the long-term population. Consider varying the '40' that appears in each one. Consider varying the rate of increase in the third possibility. Consider … .

Write a report for the farmer.

ISBN 9780170394994

ANSWERS

1 a 115 people responded to the survey.
  b 28 of the respondents chose beef as their most preferred main course at a restaurant.
  c 32% (nearest percent) of those choosing beef as their most preferred course at a restaurant were aged 31 to 43.
  d 26% (nearest percent) of those aged 31 to 43 chose beef as their preferred main course at a restaurant.
2 a The frequency table involves 376 people.
  b The frequency table involves 162 males.
  c 15% (nearest percent) of the males involved in the survey were divorced.
  d 44% (nearest percent) of the divorced people involved in the survey were males.
  e According to the survey results the percentage of males who are single (30%) is greater than the percentage of females who are single (28%).

3 a

	A	B	C	D	Totals
E	11%	13%	17%	37%	78%
F	6%	7%	7%	2%	22%
Totals	17%	20%	24%	39%	100%

b

	A	B	C	D	Totals
E	14%	16%	22%	48%	100%
F	26%	33%	32%	9%	100%

c

	A	B	C	D
E	65%	64%	71%	95%
F	35%	36%	29%	5%
Totals	100%	100%	100%	100%

For question 4, 5 and 6 the comments given here are by way of example, they are unlikely to be the only correct comments that could be made. However for each question your comments should include mention of whether the given data suggests association between the variables, as each question specifically asked for such comment.

4 a

	NSW	Queensland	Victoria	WA
Australian Rules	12%	22%	75%	72%
Rugby League	52%	47%	6%	4%
Rugby Union	21%	19%	8%	11%
Soccer	15%	12%	11%	13%
Totals	100%	100%	100%	100%

b

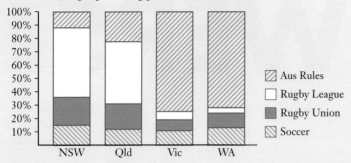

Percentages preferring particular football codes

Legend:
- Aus Rules
- Rugby League
- Rugby Union
- Soccer

c In the graph, as we move across the state columns, whilst some similarities exist there are also some very clear differences in the proportions preferring each sport, thus suggesting there is an association between the two variables, preferred football code and state lived in.

New South Wales and Queensland show similar distributions with each of these states having Rugby League as the most preferred code. 52% of NSW respondents choosing this and 47% of the Queensland respondents.

Victoria and West Australia were also similar to each other, both having Australian rules as the preferred option and both putting this well ahead of the other options. (75% of Victorian respondents and 72% of West Australian respondents).

However the 'NSW-Qld' distribution and the 'Vic-WA' distribution show considerable differences between each other. Whilst NSW and QLD had Rugby League as easily the most preferred code, Victoria and West Australia placed Rugby League as least preferred and had Australian rules as the most preferred by a considerable margin.

5 a

	Success	Failure	Totals
Drug program only	64%	36%	100%
Drug program + program A	72%	28%	100%
Drug program + program B	75%	25%	100%
Drug program + program C	71%	29%	100%

b

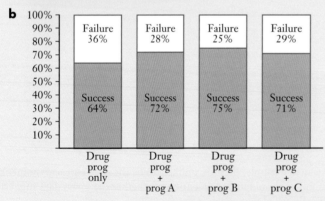

c There seems to be a reasonably consistent increase in the success rate of the program if a sufferer is on one of the programs, almost irrespective of which program, A, B or C, they are on. Hence there does seem to be an association between the success rate and whether or not a sufferer is on any one of the additional programs or not.

Sufferers who take part in one of the three additional programs, A, B or C have a 71% to 75% success rate compared to 64% proportion of success without one of the additional programs. According to the figures given in the table Program B has the greatest increase followed by A then C but the variation between the A, B and C results is not great.

(Note: We cannot necessarily assume that the additional programs are *causing* the increased success. It could be that people who opt for an additional program are the type more likely to succeed anyway. This *cause* aspect will be considered further in chapter 2.)

ISBN 9780170394994

6 a

	30 and under	31 to 40	41 to 50	51 to 60	61 and over
Male	48%	53%	52%	68%	71%
Female	52%	47%	48%	32%	29%
Totals	100%	100%	100%	100%	100%

b The proportions of male and female within each age category do not alter much as we move across the first three age categories and in each of these there is an approximately 50–50 split between males and females. Hence for these younger age categories there does not seem to be an association between age and gender balance in this workplace. However the proportions change quite dramatically in the final two categories where our previously roughly 50–50 split becomes male dominated with more than 2 males to each female in each of these last two categories. Thus there does seem to be an association between the age categories of under 50 and over 50 and gender balance in this workplace.

We are told that the organisation has in recent years introduced a workplace gender equality policy so perhaps this, maybe coupled with a change in society views, has led to the more equal gender balance in the younger, and therefore perhaps more recently appointed, employees.

7, 8 and 9 For question 7, 8 and 9 answers are not given here. Compare your answers with those of others in your class and with your teacher. Comment on the responses of others to these questions and allow them to comment on your responses.

Exercise 1B PAGE 15

1 Explanatory variable: Gender.
Response variable: Preferred film type.

2 Explanatory variable: Gender of parent.
Response variable: Preferred activity of daughter.

3 Explanatory variable: Age of mother when person was born.
Response variable: Likelihood of person developing respiratory problems.

4 Explanatory variable: Left or right handedness of mother.
Response variable: Left or right handedness of child.

5 Explanatory variable: State a person lives in.
Response variable: Preferred football code.

6 Explanatory variable: Amount of study.
Response variable: Grade in exam.

7 Explanatory variable: Population of country.
Response variable: Number of babies born in country per year.

8 Explanatory variable: How well a person sleeps at night.
Response variable: The person's grumpiness the next morning.

9 Explanatory variable: Having, or not having, the mental disorder.
Response variable: Hearing voices.

10 Explanatory variable: Gender.
Response variable: Frequency of brushing teeth.

11 Explanatory variable: Smoking classification.
Response variable: Respiratory function.

12 Explanatory variable: Age of women's mother when had first child.
Response variable: Age of women when she has her first child.

13 Explanatory variable: The time the heat has been applied for.
Response variable: The temperature of the milk.

14 Explanatory variable: Lack of sleep.
Response variable: Level of tiredness.

15 Explanatory variable: Stress levels.
Response variable: Heart disease.

16 Explanatory variable: How much alcohol the driver has consumed recently.
Response variable: How dangerous the driver is to other road users.

17 a The investigation is likely to indicate whether respiratory problems in teenage years could in any way be explained by the amount that the mother of the teenager engaged in a particular activity during pregnancy. The amount the mother engaged in the activity during pregnancy is the explanatory variable and the level of respiratory problems experienced by the child in their teenage years is the response variable.

    b The explanatory variable (level mother was engaged in the activity during pregnancy) have categories 'not at all', 'sometimes' and 'often'. These are displayed using different rows.

    c

		Level of respiratory problems in teenage years			Totals
		None	**Moderate**	**Severe**	
Engaged in activity during pregnancy	**Not at all**	74%	20%	6%	100%
	Sometimes	44%	39%	17%	100%
	Often	29%	53%	18%	100%

    d Percentage of teenagers with respiratory problems

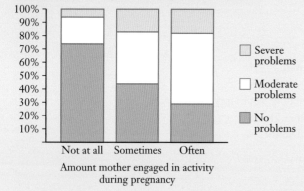

Amount mother engaged in activity during pregnancy

    e As we move across the explanatory variable categories 'not at all', 'sometimes' and 'often' the big differences between columns in the percentages in the response variable categories indicate that there is an association between the two variables.

    Further comments (given by way of example):

    The proportion of teenagers with no respiratory problems in their teenage years decreases dramatically as the engagement of the mothers in the activity during pregnancy increases. The proportion goes from 74%, for those teenagers with mothers who did not engage in the activity at all during pregnancy, to just 44% for those teenagers with mothers who engaged in the activity sometimes during pregnancy, to 29% for those teenagers with mothers who often engaged in the activity during pregnancy.

    The percentage of teenagers with 'moderate' respiratory problems increases as the engagement of the mother in the activity during pregnancy increases.

    For mothers who engage in the activity during pregnancy, the likelihood of their child having severe respiratory problems in the teenage years is roughly treble that of a child whose mother does not engage in the activity (from 6% to about 18%). However, the figures suggest that this higher figure is little changed by whether the mother's engagement in the activity was sometimes or often.

    (Note: We cannot necessarily assume that it is the mother's engagement in the activity that is causing the respiratory problems of the child. It could be that the genetic make up of mother and child is responsible for both the mother's desire to take part in the activity and the child's respiratory problems. This cause aspect will be considered further in chapter 2.)

ISBN 9780170394994

18 a We are investigating whether an individual's preferred soft drink can be explained by the age of the individual. Age is the explanatory variable and preferred choice of soft drink the response variable.

b

	Number of people nominating the drink as most preferred soft drink						Totals
	Fizzy drink	**Fruit juice**	**Coffee**	**Tea**	**Milk/ flavoured milk**	**Water**	
16 to 25	35%	25%	9%	4%	16%	11%	100%
26 to 35	24%	17%	19%	10%	12%	18%	100%
36 to 45	17%	15%	22%	12%	8%	26%	100%
46 to 55	9%	15%	25%	16%	6%	29%	100%
56 to 65	8%	16%	25%	18%	7%	26%	100%

c Whilst a graph is not specifically requested in the question it is shown below to aid interpretation.

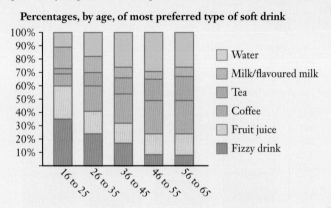

Percentages, by age, of most preferred type of soft drink

As we move across the age categories changes in the proportions of most favoured drink are very evident. For example the fizzy drink category declines from being the drink most favoured by 35% of people in the 16 to 25 age group to just 8% of people in the 56 to 65. This change in the percentages as we move across the age brackets indicates that there is an association between age and preferred choice of soft drink.

Further comments (given by way of example):

As mentioned above, the percentage of people having fizzy drinks as their most preferred choice of soft drinks declines considerably with age from 35% (16 to 25 yr age group), to 24% (26 to 35 yr age group) to 17% (36 to 45 yr age group) to 9% (46 to 55 yr age group) to 8% (56 to 65 age group).

As the age group considered increases:
> the percentage of people choosing fizzy drinks as most preferred soft drink decreases,
> the percentage of people choosing tea and coffee as most preferred soft drink increases,
> the percentage of people choosing water as most preferred soft drink increase to the 36 to 45 age group,
> after which it maintains a reasonably steady level.

With the exception of the fizzy drink decline in popularity, the last three age groups, 36 to 45, 46 to 55 and 56 to 65 have very similar percentage distributions. For these age groups water and coffee are almost equally the most preferred, at around the 25% popularity level, with water generally just ahead of coffee. Tea and Fruit Juice are next, each at around the 15% level. This is in contrast with the 16 to 25 yr age group where fizzy drink and fruit juice dominate whilst tea and coffee are least preferred.

19 a We would expect some variation in salary due to being in work part-time or full-time. Similarly we might expect some variation in salary due to age, with an older worker possibly being paid more for their experience and seniority. Limiting the survey to those in full-time employment and those aged between 30 and 40 attempts to remove from the collected data, or at least substantially reduce, salary variations due to these factors.

(However, with the reasons for salary variation being considered then the age of the workers should be considered at some stage as that aspect is likely to be a factor.)

(Part-time workers could perhaps be included if their salary details were appropriately adjusted to be comparable to a full-time worker.)

b We are investigating whether salary level can in any way be explained by blood group. Blood group is the explanatory variable and salary level the response variable. For the given table we need to calculate row percentages. (That will allow us to see if the salary level proportions change as we move across the blood groups.)

		SALARY			Totals
		Medium	**High**	**Very high**	
BLOOD GROUP	**A**	57%	35%	8%	100%
	B	51%	41%	7%	99%*
	AB	47%	40%	13%	100%
	O	56%	35%	9%	100%

*The total of 99%, rather than 100%, is simply due to rounding.

GRAPH (not essential but shown here to aid interpretation):

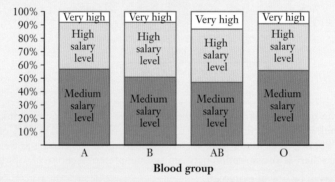

COMMENTS (given by way of example):

The percentages in the various salary levels change only a little as we move across the blood groups. The AB group proportions differ most from the others, especially at the Very high salary category but the AB group involved a very small number of people compared to the other categories (15 compared to 41, 150 and 190). Just one of these 15 in a different salary level could significantly change the AB group percentages. If we look at the two biggest blood groups, A and O, the proportions in each salary category are almost identical.

The figures suggest that for people aged between 30 and 40 working full time in this particular field there is no association between the person's blood group and their salary level.

 ISBN 9780170394994

c We are investigating whether salary level can in any way be explained by level of education. Level of education is the explanatory variable and salary level the response variable. For the given table we need to calculate column percentages. (That will allow us to see if the salary level proportions change as we move across the education categories.)

		HIGHEST LEVEL OF EDUCATION			
		School	Training diploma or certificate	Degree	Higher degree
Salary	Medium	77%	64%	56%	25%
	High	19%	31%	36%	57%
	Very high	4%	5%	8%	18%
	Totals	100%	100%	100%	100%

GRAPH (not essential but shown here to aid interpretation):

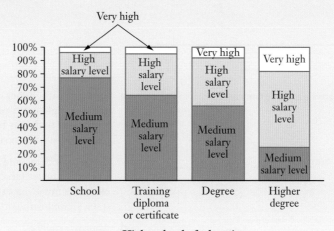

Highest level of education

COMMENTS (given by way of example):

As we move across the categories of highest level of education, from school through to higher degree:
the percentage of people on medium level salary decreases (77%, 64%, 56%, 25%),
the percentage of people on high level salary increases (19%, 31%, 36%, 57%),
the percentage of people on very high salary increases (4%, 5%, 8%, 18%).

The changing proportions indicate a strong association exists between salary levels and highest level of education.

The data suggests that for people in full time employment and aged between 30 and 40, the higher a person's level of education completed, the greater the likelihood of that person being on one of the higher salary levels.

Whilst three quarters of those who had school as their highest level of education were on the medium salary level, only one quarter of those with a higher degree were on this salary level, the other three quarters being on high or very high salaries.

20 and **21** For questions 20 and 21 answers are not given here.
Instead compare your answers with those of others in your class.
Read and comment on their report and allow them to read and comment on yours.
Ask your teacher to read and comment on your report and on the comments others have made about your report.

Note: When judging the line of best fit 'by eye' on a scattergraph there can sometimes be considerable variation between one person's judgement and another's. As you will see later in this unit there is a way in which the line of best fit can be calculated mathematically. It is this calculated line of best fit that is shown in these answers. Any predicted values you give should be based on *your* lines of best fit which should not be too far from the lines shown here.

1

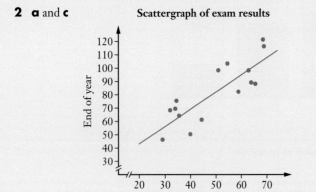

Scattergraph showing asking price of 14 cars of various ages
(Same make and model of car)

Asking price falls as age of vehicle increases.

Distribution of points suggests a linear model would be a suitable 'fit'.

Likely price for a five year old model:

Approximately $14 900.

2 a and **c**

Scattergraph of exam results

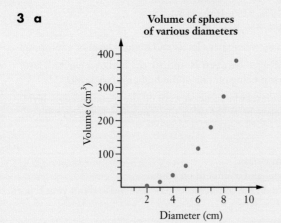

b Students with higher mid-year results tended to gain the higher end-of-year results and students with lower mid-year results tended to gain the lower end-of-year results.

A linear model would be a reasonable one to use for estimates (accepting of course that any estimates involving extrapolation would have to be viewed with particular caution).

d Approximately 82.

3 a

Volume of spheres of various diameters

b Graph indicates that a relationship exists between diameter and volume but this relationship is not linear. Hence a linear model would not be a good choice for this data.

ISBN 9780170394994

4 a

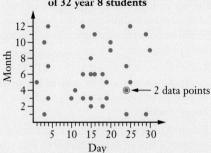

Scattergraphs of birthdays
of 32 year 8 students

2 data points

Graph indicates no relationship exists between day of birth and month of birth.

5 a

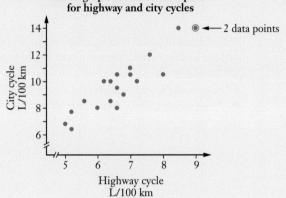

Scattergraph of fuel consumption
for highway and city cycles

2 data points

Cars that use more litres of fuel per 100 km of highway cycle also tend to use more fuel per 100 km of city cycle. The relationship between these two variables could be well represented by a linear model.

b

Scattergraph of fuel consumption
for various engine sizes

3 data points

Cars with the bigger engine size tend to use more litres per 100 km on city driving than those with the smaller engines.

Whilst the relationship between engine size and litres per 100 km could be quite well represented using a single linear model the graph perhaps better fits two straight lines, one for engine size up to about 2.8 L and then one for engine size from 3 litres upwards.

6

Scattergraph of cooking times for
joints of meat of various weights

Graph indicates that the cooking time is perfectly linearly related with the weight of the joint of meat with cooking time increasing 20 minutes for each 500 gram increase in weight.

ISBN 9780170394994

7

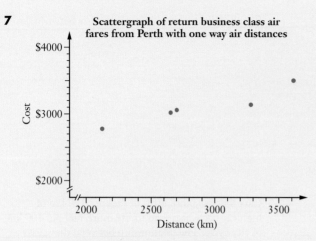

Scattergraph of return business class air
fares from Perth with one way air distances

Only a few data points so difficult to have great
confidence in any relationship suggested but the points
we have do seem to indicate a steady increase in the cost
as the distance increases. A linear model would fit the
few data points well.

b Linear fit would suggest approx $2900. This involves
interpolation so could be reliable but would like
more data points before being confident of
linear model.

c Extrapolating to 15 000 km using a linear model
suggests a fare of approx $8000 to $9000 but we are
then way beyond the known data points so could
have very little confidence in this estimate.

8

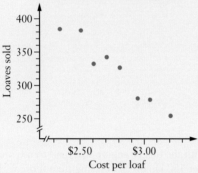

Scattergraph showing number of
loaves sold for various prices

The number of loaves sold steadily decreases as the
cost per loaf increases. A linear model would fit the
relationship well.

A linear model would suggest that for a price of $2.90
the baker needs to bake approx 300 loaves.

9

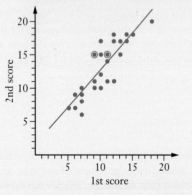

Scattergraph of memory test results

Those who scored the higher marks in the first test
tended to be the ones who also scored the higher marks
in the second test.

Predicted second score for a first score of 8 would be
approximately 10 or 11.

Miscellaneous exercise one   PAGE 27

1 a 24.5   **b** 340   **c** 112   **d** 24   **e** 36

2 The 10 means that for each one kg of extra weight suspended from the spring its length increases by 10 cm.

The 25 means that the unstretched length of the spring (i.e. the length with no weight suspended from it) is 25 cm.

3

	Coordinates of *y*-axis intercept	Gradient	Equation
A	(0, 2)	1	$y = x + 2$
B	(0, −1)	2	$y = 2x − 1$
C	(0, −3)	−2	$y = −2x − 3$
D	(0, 3)	−1	$y = −x + 3$
E	(0, 2)	0.5	$y = 0.5x + 2$
F	(0, −2)	0.5	$y = 0.5x − 2$
G	(0, 1)	−0.5	$y = −0.5x + 1$

4

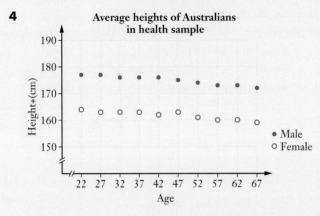

The average height of the male Australians in the sample was consistently about 13 centimetres more than the average height of the Australian females of the same age.

The graph suggests that for both males and females the trend is for the younger adults in society to have a greater average height than the older folk.

5 To investigate whether the proportions of each type of crime change much as we move across the categories of alcohol consumption we want crime proportions within each alcohol consumption category. Hence we create column percentages.

	Usual alcohol consumption			
	None	**Light**	**Medium**	**Heavy**
Assault	10%	8%	20%	32%
Fraud	24%	22%	14%	7%
Motor vehicle	52%	40%	39%	30%
Theft (not involving assault)	14%	30%	27%	32%
Totals	100%	100%	100%	101%*

*Rounding anomaly.

As we move across the categories of alcohol consumption the proportions of each crime type change quite noticeably, thus suggesting there is an association between alcohol consumption and type of crime.

The proportion of crimes that were crimes of assault was much higher (32%) amongst heavy drinkers than for non or light drinkers (10% and 8%).

The proportion of crimes that were crimes of fraud was much higher amongst non & light drinkers (24% and 22%) than among heavy drinkers (7%).

Crimes of theft had a smaller proportion amongst the non drinkers (14%) than in the other categories (~30%).

ISBN 9780170394994

1 The –1.5 suggests that on average each 1 year increase in the age of the child sees the time taken to complete the task decrease by 1.5 seconds.

The 41 suggests that a child aged 0 years would complete the task in 41 seconds. However, not only does the concept of 'age zero' not make sense in this situation, it is also outside of the age range considered, i.e. 5 to 18.

2 The 2 suggests that on average each one hour of revision sees an increase in the exam mark of 2%.

The 28 suggests that with no revision the mark gained would be 28%.

3 Scattergraph A: $y = 0.271x + 48.9$ Scattergraph B: $y = 0.283x + 26.4$

Scattergraph C: $y = -0.490x + 49.0$ Scattergraph D: $y = 0.283x + 20.7$

Scattergraph E: $y = -0.152x + 48.6$

4

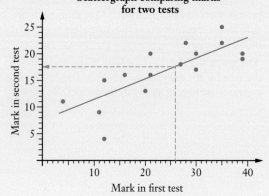

Scattergraph comparing marks for two tests

Estimated mark for absent student:

Approximately 17.5

Equation of line of best fit:

$$y = 0.39x + 7.4$$

where x is the mark in the first test and y is the mark in the second test.

5

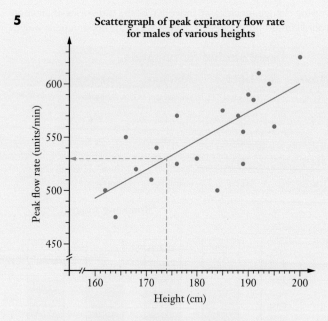

Scattergraph of peak expiratory flow rate for males of various heights

Estimated peak flow for male aged between 30 and 50 and of height 174 cm is approximately 530 units/min.

Equation of line of best fit

$$P = 2.66h + 67$$

6

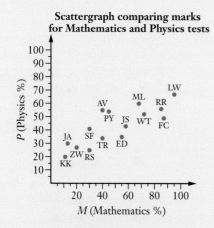

Scattergraph comparing marks for Mathematics and Physics tests

Graph indicates that in general those scoring high marks in one subject also scored high in the other, and similarly for those scoring low marks. Given the number of points involved and the reasonable linear correlation a linear model is suitable.

$$P = 0.425M + 22$$

Estimated physics mark: 49.

7 a

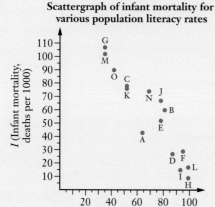

Scattergraph of infant mortality for various population literacy rates

b $I = -1.32L + 149.4$

c i 86.

The graph indicates a good linear correlation so a linear model is suitable and the prediction involves *interpolation* so should be quite a reliable prediction.

iii 130.

This time *extrapolation* is involved with a literacy rate of 15% being well below the given literacy rates (for which the lowest is 35%). Hence prediction should not be relied upon.

Exercise 2B   PAGE 46

1 No we cannot conclude that the reading of fashion magazines is causing the eating disorder. The fact that there is an observed association between the number of hours spent reading teen fashion magazines and the chance of having an eating disorder does not mean that one causes (or doesn't cause) the other.

What might be more likely to be happening here is that a person's level of concern about their own body image could be a confounding, or lurking, variable. If a person is overly concerned with their own appearance and body image this concern could drive them to look through teen fashion magazines to view pictures of other young people, and could also drive them into eating disorders as they seek their desired look.

2 No we cannot conclude that the duration of employment of the individual is causing the decline in company profits. The fact that there is an observed association between two variables does not mean that one necessarily causes the other. It is more likely in this situation that it is purely coincidental that the individual happened to begin employment at a time when company profits started their decline.

3 No we cannot conclude that it is the extra time at school that is causing the rise in the number of school leavers with attention deficit disorder. The fact that there is an observed association between two variables does not mean that one necessarily causes the other.

The confounding variable could well be time. With time the population will increase and so we might expect the number of cases of particular disorders to increase. As society evolves over time greater importance is attached to education, hence the increasing time at school and the greater number of years when learning order diagnoses could occur. Also, with time, greater awareness of learning disorders and better methods of diagnosing them develop, hence the rise in the number diagnosed with attention deficit disorder.

4 a r = 0.74   **b** $y = 0.346x + 5.48$ **c** 15.2   extrapolation

5 a r = −0.87   **b** $y = -0.640x + 17.74$ **c** 13.9   interpolation

6 a r = –0.84   **b** $y = -0.385x + 18.87$

  c i 15.0   interpolation   **ii** 3.5   extrapolation

7 a r = 0.88   **b** $y = 0.365x + 3.83$

  c i 18.4   extrapolation   **ii** 10.0   interpolation

8 a r = 0.70   **b** $y = 0.095x + 2.56$ **c** 3.6   extrapolation

9 a r = –0.88   **b** $y = -0.663x + 46.7$

  c i 32.1   interpolation   **ii** 15.5   interpolation

10 a r = 0.87   **b** $y = 37.8x + 239.2$

  c i 432   interpolation   **ii** 587   extrapolation   **iii** 727   extrapolation

11 a r = 1   **b** $y = 0.8x - 11$

  c i 57   interpolation   **ii** 21   interpolation   **iii** 149   extrapolation

12 a $q = -0.655p + 34.22$ **b** $p = -1.199q + 46.24$

  c i 14.6   **ii** 16.3

13 a $q = 0.2p + 12$ **b** $p = 5q - 60$

  c i 21.6   **ii** 100

14 a –0.73   **b** $N = 399 - 8.38T$

  c i 181   Should be fairly reliable given r value and interpolation involved.

    ii 47   Extrapolation involved so validity is questionable.

15 a 0.83   **b** $N = 37.8 + 9.8T$

 c i 175   Should be fairly reliable given r value and interpolation involved.

    ii 8   Extrapolation involved so validity is questionable.

16 a 0.85   **b** End of year = $19.7 + 1.16 \times$ mid year   **c** 93

17 a –0.97, $S = 35.3 - 0.38C$ **b** ~ 79%

18 a 0.86, $L = 52.1 + 0.20S$ **b i** 64 years   **ii** 68 years

  c No we cannot conclude that a high life expectancy is due to having access to good sanitation. The data suggests there is a correlation between the two but we cannot conclude that one is due to the other. I.e. correlation does not allow us to assume cause and effect.

Exercise 2C   PAGE 55

1 a $y = -0.366x + 87.467$, r = –0.645   **b** $y = -0.550x + 99.144$, r = –0.957

2 a

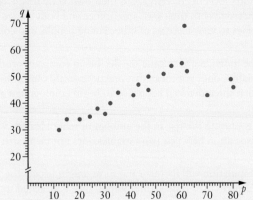

b $q = 0.491p + 24.482$

c 43.1

d Predicting q for $p = 75$ involves extrapolating beyond the values that contributed to the regression line, hence unwise. Furthermore we have data points for this sort of p value that already suggest that our regression line may not be at all suitable for these values of p, again suggesting that it would not be wise to use our regression line to predict a q value for $p = 75$.

3 Table 1, Graph C.   Table 2, Graph D.   Table 3, Graph E.

  Table 4, Graph A.   Table 5, Graph F.   Table 6, Graph B.

4 According to the given information, approximately 64% ($0.802^2 \approx 0.64$) of the variation in the weights of men of the particular age can be explained by the variation in their heights.

5 Approximately 53% ($0.73^2 \approx 0.53$) of the variation in reading age can be explained by the variation in foot length.

The given information does not mean that increasing the length of a person's foot causes their reading age to increase. Correlation shows association between variables but does not prove that the association is causal. Amongst primary school children we would expect reading age to generally increase with increasing foot size because the increased foot size will often mean an older child is being considered, and we would expect the older children to generally have higher reading ages. Hence the association is not a surprise but it does not prove cause and effect. In this case the 'lurking variable', or 'confounding variable', is the age of the primary school child.

6 (Graph not shown here.)

r = 0.51   (correct to two decimal places),

r^2 = 0.26 (correct to two decimal places).

The value of r informs us that *x* and *y* have a moderate positive linear association.

The value of r^2 informs us that 26% of the variation in the variables is explained by the linear association between them.

7 Viewing the scattergraph on a calculator or computer confirms that there seems to be an association between *units of fertiliser added* and *mean height of the seedlings after six weeks* that could well suit a linear model. (This is further confirmed by a correlation coefficient of 0.9736 indicating a very strong linear correlation.)

With the correlation coefficient, r, equal to 0.9736, $r^2 \approx 0.95$. Hence approximately 95% of the variation in seedling height can be accounted for by the variation in the number of units of fertiliser added.

8 a

t days	0	1	2	3	4	5	6	7	8	9	10
Residual	69	39	−37	−25	−36	−45	−35	−10	9	30	42

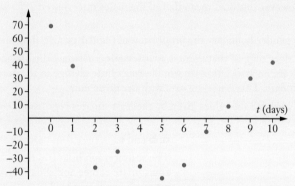

b Residuals are not random. Each end of the table is giving positive residuals, negative residuals occurring in the middle of the table. This suggests that using this linear model to predict *P* for higher *t* values will increasingly underestimate the *P* value.

9 a

h	1	2	3	4	5	6	7	8	9	10
Residual	245	82	−41	−124	−162	−165	−123	−41	81	243

b Residuals are not random. Suggests linear model may not be the most appropriate.

c $a = 20.4, b = 8.5, c = -13.9$.

d r ≈ 4700

Viewing the best fit quadratic on a calculator shows an extremely good fit to the given data points. However the prediction involves extrapolation so, without further knowledge, the predicted value should not be regarded as being particularly reliable, especially as we are extrapolating quite a long way from measured *h* values.

1 No we would not be correct to conclude that the firemen were causing the damage. Correlation does not prove causation. The fact that A and B are correlated does not of itself prove any causal relationship between them. In the given situation a more sensible explanation is that the correlation is due to some other factor, such as the size of the fire. The larger the fire the more damage we would expect to be done and the more firemen likely to be in attendance. Thus size of fire is likely to be the lurking, or confounding, variable.

2 We wish to see if the proportion who are regular cyclists changes as we move across the gender boundary. Are the proportions in any way explained by gender? Thus we need column percentages. Or, using the rule, the explanatory variable is gender and this uses columns to display its categories so we want column percentages.

	Male	**Female**
Regular cyclist	28%	16%
Not a regular cyclist	72%	84%
Totals	100%	100%

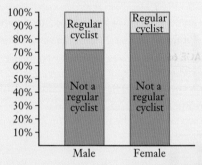

For the group interviewed the majority of both the males and the females were not regular cyclists (72% of the males and 84% of the females). However this does mean that of the males interviewed 28% were regular cyclists compared to 16% of the females.

Thus as we move across the gender boundary the proportion of regular cyclists does show quite a change.

The research data collected does suggest that there is an association between the likelihood of someone being a regular cyclist and gender. If there were no association we would expect little change in the percentage of regular cyclists as we move across the gender boundary. This is not the case with the given data.

3 (The predicted values given here are the values given by the least squares regression line. Your 'eyeballed' values should be reasonably close to these values.)

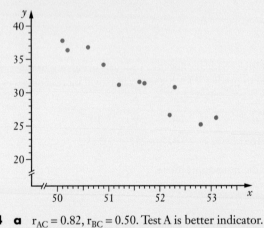

a, b and **c:**

The scattergraph indicates that a linear model would be an appropriate choice and should give reliable predictions when interpolation is involved.

a When $x = 50.8$ the linear model suggests $y \approx 34.5$. The suitability of the model and the fact that interpolation is involved means that this predicted y value should be very reliable.

b When $x = 53.2$ the linear model suggests $y \approx 24.7$. This prediction involves extrapolation but 53.2 is not far beyond the given x values so the prediction should still be viewed as quite reliable.

c When $x = 58$ the linear model suggests $y \approx 5$ but $x = 58$ involves extrapolation far beyond the given x values. Hence the predicted y value should not be considered at all reliable.

4 a $r_{AC} = 0.82$, $r_{BC} = 0.50$. Test A is better indicator.   **b** 70   **c** 1, 3, 8, 9, 11, 15.

5 a The test B marks were generally higher than the test A marks, with B having a mean score of 37.77 and a median of 39, and A a mean score of 32.43 and a median of 31. Twenty one of the thirty students achieved a higher mark in test B than they did in test A, two students scored the same and seven scored more in A than in B.

Both sets of scores had a low score of 17 and a high score of 49 and hence the sets of scores had the same range (32). However in set B, two scores lie noticeably further away from the rest so whilst the range of the two sets is the same, other measures of spread indicate that the set B scores are less spread out than the set A scores. The interquartile range is 15 for set A and 9 for set B and standard deviation is 9.29 for set A and 7.20 for set B (9.45 and 7.32 if using σ_{n-1}).

If the two outliers are removed from the test B scores this difference in spread is even more noticeable.

b The shape of the scattergraph suggests that linear regression is appropriate and the score A mark of 27 achieved by the 31st student means interpolation will be involved, further justifying the validity of regression analysis. Cropping the data so that the two outlying points are excluded, and hence do not unduly influence the regression line, gives a regression line:

$$\text{Score B} = 0.497 \times \text{Score A} + 22.70$$

and a predicted mark of 36.1 in B.

With 28 points involved, good linear correlation (r = 0.883) and interpolation used, the estimate is likely to be reasonably valid. However it is only an estimate and, as two students did show, it is quite possible to score a lower mark in test B than the test A mark would generally suggest.

(With the two outlying scores included the regression line is

$$\text{Score B} = 0.611 \times \text{Score A} + 17.94 \text{ to give a predicted mark of 34.4.})$$

Exercise 3A   PAGE 65

1	13, 15.	**2**	320, 640.	**3**	14, 16.	**4**	G, H.
5	S, V.	**6**	VII, VIII.	**7**	22, 25.	**8**	J, A.
9	13, 21.	**10**	49, 64.	**11**	17, 19.	**12**	28, 36.
13	0.00001, 0.000001.	**14**	64, 128.	**15**	729, 2187.	**16**	79, 72.
17	3412, 4123.	**18**	73, 84.	**19**	0.015625, 0.0078125.	**20**	$0.\overline{142857}$, 0.125.
21	127, 255.	**22**	−1, −3.	**23**	364.5, 546.75.		
24	823543 (= 7^7), 16777216 (= 8^8).						

Exercise 3B   PAGE 67

1	14	**2**	17	**3**	31	**4**	42	**5**	44	**6**	28
7	39	**8**	23	**9**	32	**10**	3	**11**	3	**12**	3
13	7	**14**	255	**15**	262	**16**	21	**17**	30	**18**	157
19	23	**20**	16	**21**	511	**22**	4	**23**	8	**24**	16
25	26	**26**	15	**27**	17	**28**	32	**29**	3	**30**	5
31	7	**32**	65	**33**	1	**34**	3	**35**	2	**36**	3
37	5	**38**	42	**39**	0	**40**	55	**41**	1	**42**	7
43	4	**44**	7	**45**	11	**46**	94	**47**	0	**48**	123

Exercise 3C   PAGE 73

1 a I   **b** A   **c** B   **d** D   **e** G   **f** K
   g H   **h** D   **i** F

2 11, 16, 21, 26, 31, 36.   **3** 4, 3, 1, −3, −11, −27.   **4** 3, 7, 11, 15, 19.

5 1, 4, 7, 10, 13.   **6** 2, 5, 11, 23, 47.   **7** 2, 5, 14, 41, 122.

8 1, 3, 13, 63, 313.

9 $T_1 = 128, T_2 = 160, T_3 = 200, T_4 = 250, T_{n+1} = 1.25 \times T_n$.

10 $T_1 = 1000, T_2 = 1200, T_3 = 1440, T_4 = 1728, T_{n+1} = 1.2 \times T_n$.

11 $T_1 = 2000, T_2 = 1700, T_3 = 1445, T_4 = 1228.25, T_{n+1} = 0.85 \times T_n$.

12 $T_1 = 5, T_{n+1} = T_n + 4$.   **13** $T_1 = 26, T_{n+1} = T_n − 3$.   **14** $T_1 = 6, T_{n+1} = 2T_n + 1$.

15 $a = 4, T_3 = 24$.   **16** 15, 23, 31, 39, 47, 55.   **17** 10, 26, 74, 218, 650, 1946.

18 12, 23, 45, 89, 177, 353.   **19** 1, 4, 5, 9, 14.   **20** 7, 7, 14, 21, 35.

21 3, 4, 2, 6, −2.   **22** 1, 2, 3, 6, 11, 20.   **23** $a = 5, b = 3, T_6 = 5468$.

24 $a = 2, b = −3, T_6 = 67$.   **25** 69, 56, 47, 41, 37.25.   **26** 1, −2, −4, 8, 16.

27 3, 6, −12, −24, 48.   **28** 3, 9, 144, 18000, 23328000.

29 a

		Stacking cans			
T_1	T_2	T_3	T_4	T_5	T_6
1	3	6	10	15	21

b IV

Spreadsheets   PAGE 76

1

	A	B	C	D
1	1	6	5	1
2	2	7	10	4
3	3	8	15	9
4	4	9	20	16
5	5	10	25	25
6	6	11	30	36
7	7	12	35	49
8	8	13	40	64

2 Square

3 Yes:

A	B	C	D
n	$n + 5$	$5n$	?

$? = n(n + 5) - 5n$
$= n^2 + 5n - 5n$
$= n^2$, as required.

Miscellaneous exercise three   PAGE 76

1 a 1093, 3280.   **b** 343, 512.   **c** 262.144, 209.7152.   **d** −200, −400.   **e** $\frac{7}{8}$, 1.

2 1, 2, 9, 58.

3 120, 58, 27, 11.5.

4 The 3.1 suggests that in general, if the consumption of the foodstuff in a country goes up by 1 kilogram per person per year we would expect to see the death rate due to heart disease for that country increase by approximately 3.1 deaths per year per 100 000 of population.

The 90.9 suggests that if a country has a zero consumption of the foodstuff the death rate due to heart disease would be approximately 90.9 deaths per year per 100 000 of population. (However, zero consumption of the foodstuff is outside the range of consumption figures used to form the line of best fit so this suggested death rate due to heart disease figure should not be relied upon.)

5 $a = 11$, $T_3 = 278$.

6 20, 42, 40, 4, −36.

7 The student's conclusion is incorrect. The data is not best suited to linear modelling. Had the student viewed a scattergraph of the data, see right, it should have been apparent that a linear model was probably not the most appropriate but that an exponential or perhaps a polynomial function would have been better.

A look at the residuals or a residual plot, see below (residuals calculated using equation given in question), should similarly have alerted the student to this fact as both show there is a pattern to the residuals and for linear modelling to be appropriate this should not be the case.

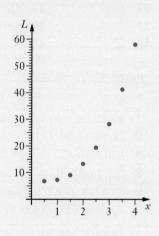

x	0.5	1.0	1.5	2.0	2.5	3.0	3.5	4.0
Residual	8.55	2	−3.25	−6.1	−7.05	−5.3	0.65	10.4

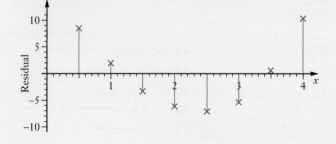

ISBN 9780170394994

8 a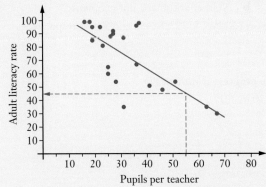

b No. Data suggests there is a correlation between literacy rate and pupils per teacher but cannot conclude one is due to the other.

c Approx 45%.

9 a For the line $y - \bar{y} = \dfrac{s_{xy}}{s_x^2}(x - \bar{x})$ to pass through $(\bar{x}, \bar{y})$ the values $x = \bar{x}$ and $y = \bar{y}$ must 'fit' the equation.

Substituting $x = \bar{x}$ gives $\qquad y - \bar{y} = \dfrac{s_{xy}}{s_x^2}(\bar{x} - \bar{x})$

Hence $\qquad\qquad\qquad y - \bar{y} = 0$

i.e. $\qquad\qquad\qquad\qquad y = \bar{y}$

Thus the least squares regression line passes through $(\bar{x}, \bar{y})$.

b Because it is the line that *minimises* the sum of the *squares* of the residuals.

10 Allow others to read and comment upon your report and similarly you read and comment upon the reports of others.

Exercise 4A   PAGE 87

1 $T_1 = 6, T_{n+1} = T_n + 4$.

2 $T_1 = 28, T_{n+1} = T_n - 2$.

3 $T_1 = 5, T_{n+1} = T_n + 10$.

4 $T_1 = 7.5, T_{n+1} = T_n + 2.5$.

5 $T_1 = 100, T_{n+1} = T_n - 11$.

6 $T_1 = 6, T_n = 2T_{n-1}$.

7 $T_1 = 0.375, T_n = 4T_{n-1}$.

8 $T_1 = 384, T_n = 0.25T_{n-1}$.

9 $T_1 = 50, T_n = 3T_{n-1}$.

10 $T_1 = 1000, T_n = 1.1T_{n-1}$.

11 Neither arithmetic nor geometric.

12 Geometric.

13 Arithmetic.

14 Arithmetic.

15 Neither arithmetic nor geometric.

16 Geometric.

17 Geometric.

18 Arithmetic.

19 Neither arithmetic nor geometric.

20 Neither arithmetic nor geometric.

21 Arithmetic.

22 Geometric.

23 $T_1 = 8, T_2 = 11, T_3 = 14, T_4 = 17, T_{n+1} = T_n + 3$.

24 $T_1 = 100, T_2 = 97, T_3 = 94, T_4 = 91, T_{n+1} = T_n - 3$.

25 $T_1 = 11, T_2 = 22, T_3 = 44, T_4 = 88, T_{n+1} = 2T_n$.

26 $T_1 = 2048, T_2 = 1024, T_3 = 512, T_4 = 256, T_{n+1} = 0.5T_n$.

27 b $N_{n+1} = N_n + 800$

28 a The sequence is a geometric progression.

b The next three terms after the first are 550, 605 and 665.5.

29 a The sequence is a geometric progression.

b The next three terms after the first are 1250, 1562.5 and 1953.125.

30 a The sequence is a geometric progression.

b The next three terms after the first are 21 600, 19 440 and 17 496.

31 a $T_1 = 3, T_{n+1} = T_n + 5$.

b The sequence is arithmetic.

32 a $T_1 = 1.5, T_{n+1} = T_n \times 2$.

b The sequence is geometric.

33 a $T_1 = 4, T_2 = 9, T_3 = 16, T_4 = 25, T_5 = 36$.   **b** Neither arithmetic nor geometric.

34 a After 1 year, 2 years, 3 years and 4 years the account is worth $1296, $1392, $1488 and $1584 respectively.

    b The amounts are in arithmetic progression.

    c $T_1 = \$1200, T_{n+1} = T_n + \$96$.

35 $T_1 = \$45\,000, T_{n+1} = T_n + \$1500$. The terms of the sequence progress arithmetically.

36 $68\,000 in 2014, $71\,400 in 2015, $74\,970 in 2016, $78\,718.50 in 2017.

    $T_1 = \$68\,000, T_{n+1} = 1.05T_n$.

37 $T_1 = \$1500, T_{n+1} = 1.08T_n$.

38 a $21.00   **b** $37.90   **c** $54.80

39 $124\,000, $110\,000, $96\,000.

40 $22\,750, $18\,500, $14\,250.

    After 70\,000 copies in its first year the value of the copier will be $26\,405.

41 $T_1 = \$36\,000, T_{n+1} = 0.85T_n$.

Exercise 4B   PAGE 97

1 $T_{100} = 506$ **2** $T_{100} = 289$ **3** $T_{100} = 815$

4 $T_{100} = -120$ **5** $T_{25} = 5 \times 2^{24}$ **6** $T_{25} = 1.5 \times 4^{24}$

7 $T_{25} = 8 \times 3^{24}$ **8** $T_{25} = 11 \times 2^{24}$ **9** $T_{28} = 223$

10 $T_{20} = 3\,495\,265$ **11** $T_{19} = 774\,840\,977$ **12** $T_{45} = 6$

13 $T_1 = 48, T_{n+1} = T_n + 3$. Julie successfully completes 90 items on the 15th day.

14 Substituting y for T_n and x for n the rule $T_n = a + (n-1)d$ becomes $y = dx + (a-d)$.

    This is the equation of a straight line with gradient d, cutting the y-axis at $(0, a-d)$.

15 Substituting y for T_n and x for n the rule $T_n = ar^{n-1}$ becomes $y = ar^{x-1}$, i.e. $y = \dfrac{a}{r}r^x$.

    An exponential function cutting the y-axis at $\left(0, \dfrac{a}{r}\right)$.

16 As $n \to \infty$ the 'nd' term in the expression $a + (n-1)d$ will dominate.

    Thus as $n \to \infty$, T_n will be increasingly large and positive if $d > 0$
                                    and increasingly large and negative if $d < 0$.

17 As $n \to \infty$ the n in the expression ar^{n-1} will dominate.

    Thus as $n \to \infty$,   if $r > 1$, T_n will become increasingly large, either positively or negatively dependent on the sign of the constant a.

                    if $r < -1$, T_n will become increasingly large, alternating between large negative and large positive.

               if $-1 < r < 1$, T_n will become smaller and smaller, maintaining the sign of the constant a if r is positive and alternating between small positive and small negative if r is negative. I.e., if $-1 < r < 1$, as $n \to \infty$, $T_n \to 0$.

18 The first four terms are 8, 11, 14, 17. The 50th term is 155. The 100th term is 305.

19 The first four terms are 100, 97, 94, 91. The 50th term is –47. The 100th term is –197.

20 11, 22, 44, 88, 180\,224, 184\,549\,376.

21 2048, 1024, 512, 256, 0.0625.

22 a $T_n = 9 + (n-1) \times 6$, i.e. $T_n = 6n + 3$.   **b** $T_n = 7 + (n-1) \times 1.5$, i.e. $T_n = 1.5n + 5.5$.

23 a $T_n = 3 \times 2^{n-1}$ **b** $T_n = 100 \times 1.1^{n-1}$

24 a 856   **b** 3495   **c** The 142\,858th term.

25 a 126\,953.125   **b** The 14th term.

26 a 844\,700   **b** The 60th term.

ISBN 9780170394994

27 1, 8, 27, 64, neither.

28 a 64   **b** 7

29 a 1850   **b** 2000

30 a 7 971 615   **b** 5

31 a −1 835 008   **b** 7

32 The amount in the account at the end of ten years is $8635.70.

33 a 12 800   **b** 10 800   **c**

34 a $48 000   **b** $59 000

35 Approximately 61 million, assuming annual growth rate for the given years continues.

36 Approximately 6000.

37 a 1.2 m   **b** Approximately 9 cm.

38 a $3 130 000   **b** $3 920 000

     c Approximately fifteen and a half years.

39 Just after the end of the 22nd year, i.e. early in the 23rd year.

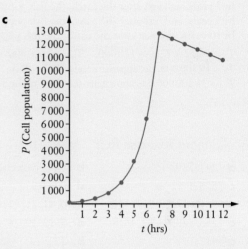

Exercise 4C   PAGE 102

1 $T_1 = 10$, $T_2 = 23$, $T_3 = 49$, $T_4 = 101$.

2 $T_1 = 200$, $T_2 = 450$, $T_3 = 1200$, $T_4 = 3450$.

3 $T_1 = 200$, $T_2 = 150$, $T_3 = 50$, $T_4 = -150$.

4   Sequence A   Sequence B   Sequence C

   a $T_1 = 5$ **a** $T_1 = 30$ **a** $T_1 = 90$

       $T_2 = 11$ $T_2 = 30$ $T_2 = 85$

       $T_3 = 23$ $T_3 = 30$ $T_3 = 75$

       $T_4 = 47$ $T_4 = 30$ $T_4 = 55$

       $T_5 = 95$ $T_5 = 30$ $T_5 = 15$

   b   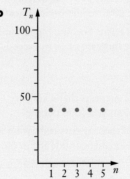   **b** 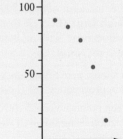

   c   Increasing sequence.   **c**   Steady state sequence.   **c**   Decreasing sequence.

5 The terms increase towards a value of 50 000, getting closer and closer to 50 000 as n becomes increasingly large.

Multiplication by 0.96 produces a 4% decrease but we also add 2000. With 2000 being greater than 4% of 35 000 the terms will increase in size but eventually the 4% decrease will almost balance the 2000 increase. The long term steady state referred to in the question will result as the decrease, 4% of T_n, approaches the increase, 2000, i.e. as T_n approaches 50 000.

6 In 1 year, and just after the catch, the fish population would be approximately 1940.
In 2 years, and just after the catch, the fish population would be approximately 1873.
In 5 years, and just after the catch, the fish population would be approximately 1619.
In 10 years, and just after the catch, the fish population would be approximately 947.

7 $T_n = 1.012 \times T_{n-1} + 120\,000, \quad T_0 = 17\,600\,000.$
$T_5 \approx 19\,300\,000$, i.e. approximately 19.3 million.
$T_{10} \approx 21\,100\,000$, i.e. approximately 21.1 million.

Miscellaneous exercise four   PAGE 103

1 a 243, 729.
b O, U (vowels) or M, Q (alphabetical, jumping three letters each time).
c S, S.
d 0.5, 0.25.

2 a 7, 17, 37, 77, 157, 317.
b 6, 11, 21, 41, 81, 161.
c 256, 128, 64, 32, 16, 8.
d 5, 12, 19, 26, 33, 40.

3 a 43
b 22
c 6
d 8
e $a = 4, b = -1, T_9 = 43\,691, T_{10} = 174\,763.$

4 157
5 0.1875

6 The 0.9 suggests that in general, if we take two people from this group, for each 1 cm difference in height we would expect to find the taller person weighing approx 0.9 kg more than the shorter.

7 $r_{xy} = 0.86, y = 29.5 + 0.532x$
a 46.5
b 61.4

A scattergraph of the data, and the value of r for the ten data points, both indicate that it is valid to use linear regression to predict. However, we must view the part **b** prediction with great caution as extrapolation is involved with $x = 60$ quite a way beyond the x values ($22 \leq x \leq 49$) in the given data.

8 a From the first term value of 5000 subsequent terms decrease, getting closer and closer to 4000 as n gets bigger and bigger. (Because as T_n gets closer to 4000 the 20% reduction is almost exactly balancing the 800 that is added.)

b From the first term value of 5000 subsequent terms increase, getting closer and closer to 8000 as n gets bigger and bigger. (Because as T_n gets closer to 8000 the 20% reduction is almost exactly balancing the 1600 that is added.)

c From the first term value of 5000 subsequent terms become bigger and bigger with $T_n \to \infty$ as n becomes increasingly large. (Because from the start the increase of 20% exceeds the 800 we are taking away so subsequent terms will increase 'without bound'.)

9 $T_{n+1} = 1.08 \times T_n - \$200, \quad T_{10} = \$5738.39.$

10 (Graph not shown here.) Applying a linear model suggests that a country with 285 motor vehicles per 1000 people would have approximately 385 (nearest 5) televisions per 1000 people.

With eighteen points involved, the distribution of those points indicating a linear model to be appropriate, r = 0.936 and interpolation being involved we can have reasonable confidence in our predicted value.

Approximately 88% ($r^2 = 0.876$) of the variation in numbers of televisions can be explained by the variation in the number of motor vehicles.

	Never	**Once per week**	**Twice per week**	**Three times per week**	**Four times per week**	**Five times per week**
Male	24%	10%	12%	18%	20%	16%
Female	24%	8%	14%	15%	21%	18%

As we move across the gender categories the proportions in each frequency of engagement in sport or exercise change very little. For both males and females the once a week attendance produced the lowest percentages of 10% and 8% respectively. For both gender categories 24% of those surveyed said that in an average week they never engage in sport or exercise.

The figures suggest that there is no association between gender and the frequency with which an adult engages in sport or exercise.

Exercise 5A   PAGE 109

1 **a** From Kellerberrin to Merredin is 58 km.
   b From Northam to Coolgardie is 461 km.
   c The entire journey would take 7 hours and 15 minutes.
2 ABDF 237 km, ABCDF 264 km, ACDF 213 km, ACEF 225 km, AEF 215 km.
3 The cheapest the system can be done for is $3990.
4 Vertices 5, Edges 5, Faces 2.   5 Vertices 5, Edges 8, Faces 5.   6 Vertices 8, Edges 8, Faces 2.
7 Vertices 2, Edges 1, Faces 1.   8 Vertices 5, Edges 6, Faces 3.   9 Vertices 5, Edges 8, Faces 5.
10 Vertices 5, Edges 8, Faces 5.   11 Vertices 7, Edges 10, Faces 5.   12 Vertices 1, Edges 2, Faces 3.
13 Vertices 5, Edges 7, Faces 4.   14 Vertices 7, Edges 8, Faces 3.   15 Vertices 7, Edges 9, Faces 4.
16 Vertices 5, Edges 9, Faces 6.   17 Vertices 7, Edges 12, Faces 7.   18 Vertices 6, Edges 9, Faces 5.
19 Vertices 6, Edges 9, Faces 5.   20 Vertices 10, Edges 13, Faces 5.   21 Vertices 12, Edges 16, Faces 6.

Exercise 5B   PAGE 124

1 **a** Connected   **b** Connected   **c** Disconnected   **d** Disconnected   **e** Connected
2 **c** and **e** are simple graphs.
   a is not a simple graph because it contains multiple edges.   **b** is not a simple graph because it contains a loop.
   d is not a simple graph because it is directed.   **f** is not a simple graph because it is weighted.
3 **a** Length 5   **b** Length 3   **c** Length 6
4 **a** Digraph.   A might like B but B does not necessarily like A. Hence the need to show direction.
   b Digraph.   Just because A is aware of the existence of B it does not necessarily follow that B is aware of the existence of A. For example suppose B is someone famous. A may well be aware of the existence of B without B being aware of the existence of A. Hence the need for direction.
   c Undirected.   No need to show direction because if A is accepted as a Facebook friend of B then B is automatically a Facebook friend of A.
   d Undirected.   No need to show direction. If A is related to B then B must be related to A.
   e Undirected.   No need to show direction. If A has been to the movies with B then B must have been to the movies with A.
   f Undirected.   No need to show direction. If A lives within 5 km of B then B must live within 5 km of A.
   g Digraph.   Because A has the phone number of B it does not necessarily follow that B must have the phone number of A. Hence the need to show direction.

ISBN 9780170394994

5 a If the edge CE is removed the graph remains connected, see diagram on the right. Hence CE is not a bridge.

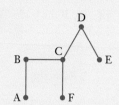

b In the connected network given in the question, edges AB, BC and CF are bridges because, in each case, removing the edge leaves the network unconnected, as shown below.

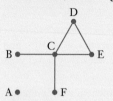

Edge AB removed
Graph unconnected

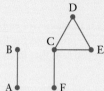

Edge BC removed
Graph unconnected

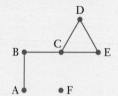

Edge CF removed
Graph unconnected

6 Only **c** and **d** are complete graphs.

7

8 a Of the given sequences of vertices the walks are: ABCDE, BCD, ABDEF, ABDEFBC, BDEFB.

b ABCDE: A path.   BCD: A path.   ABDEF: A path.   ABDEFBC: A trail.   BDEFB: A cycle

9 a

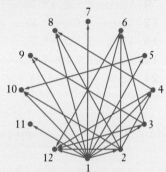

b

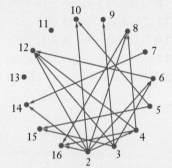

10

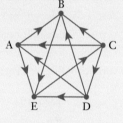

Exercise 5C   PAGE 130

1 The network has 6 edges.   **2** The network has 4 vertices.   **3** The network has 9 faces.

4 a A cube has 8 vertices, 6 faces, 12 edges. Euler's rule is obeyed.

b A regular tetrahedron has 4 vertices, 4 faces, 6 edges. Euler's rule is obeyed.

c A regular octahedron has 6 vertices, 8 faces, 12 edges. Euler's rule is obeyed.

5 Only graph **c** is a subgraph of the given graph.

Graph **a** is not because it contains the edge FE, which is not in the original.

Graph **b** is not because it contains the edge CD, which is not in the original.

6

PB   BK   MO   FS

Arsenal   Everton   Liverpool   Manchester City   Manchester United

7 a If the broken line were to be added to the graph, as shown on the right, the graph would be complete because every vertex would be joined to every other vertex by an edge. However without this line there are two vertices that are not joined by an edge and hence the graph as presented in the question is not complete.

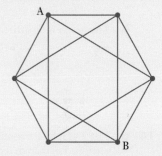

b

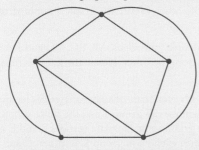

8 a The graph is redrawn on the right with two vertices labelled A and B.

These two vertices are not joined by a single edge.

Hence the graph is not a complete graph.

(The vertices labelled A and B are not the only pair of vertices that could be chosen to make this point.)

b

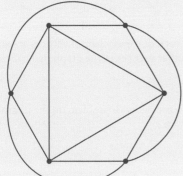

9

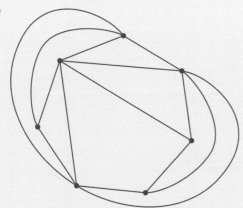

10 {A, D, F} and {B, C, E, G}

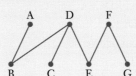

11 No the given graph is not a bipartite graph.

Suppose we were to split the vertices into two groups. Vertex A would have to be in one of the groups and so, with A having an edge to vertex B and an edge to vertex C, B and C must be in the other group if A is not to be joined to a vertex in its own group. However B and C would then be in the same group as each other and are joined by an edge. Hence not every edge would join a vertex from one group to a vertex of the other. Hence the graph is not bipartite.

12 Yes the given graph is a bipartite graph.

Divide the vertices into two groups, {A, C} and {B, D}. The given graph involve four edges, AB, BC, CD and DA. Each one of these joins a vertex from one of these two groups to a vertex from the other. Hence the graph is bipartite.

13 a The graph is not bipartite.

To be bipartite we must be able to split the vertices into two groups with every pair of adjacent vertices having one vertex from each group. Suppose that whatever group we put vertex A in we call group 1 and the other group we call group 2. An edge joins A to B and an edge joins A to E. Hence for the graph to be bipartite vertices B and E must be in group 2. But then C, being joined to B, and D, being joined to E, must each be in group 1. But then C and D are joined but in the same group. Hence the graph cannot be bipartite.

b The graph is bipartite.

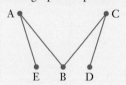

c The graph is bipartite.

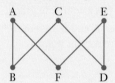

14 a The 3 graphs are complete graphs because each is a simple graph and in each one every pair of vertices are adjacent vertices. I.e. there is a path from each vertex in the graph to each of the other vertices in the graph that passes along just one edge.

b No it cannot be done for the other two graphs.

15 Compare your answer with that of others in your class and read about this classic network problem on the internet by searching for *The Water, Gas and Electricity Problem* or *The Three Utilities Problem*.

Exercise 5D   PAGE 137

1 A degree 3.   B degree 4.   C degree 1.   D degree 3.   E degree 3.   F degree 4.
  G degree 1.   H degree 3.   I degree 4.   J degree 3.   K degree 2.   L degree 5.
  M degree 4.   N degree 2.

2 A in-degree 2,   out-degree 2.   B in-degree 2,   out-degree 1.
  C in-degree 1,   out-degree 2.   D in-degree 1,   out-degree 2.
  E in-degree 2,   out-degree 2.   F in-degree 3,   out-degree 2.
  G in-degree 2,   out-degree 2.   H in-degree 1,   out-degree 1.
  I in-degree 1,   out-degree 2.   J in-degree 1,   out-degree 1.
  K in-degree 2,   out-degree 1.

3 a Semi-Eulerian.   **b** Neither Eulerian nor semi-Eulerian.
  c Eulerian.   **d** Semi-Eulerian.
  e Semi-Eulerian.   **f** Eulerian.

4 a Many possible correct answers, one of which is ABCAFEDCEA.
  b Many possible correct answers, one of which is ABCDEFA.

5 a The graph is neither Eulerian nor semi-Eulerian. (Four odd vertices.)
  b The graph is semi-Hamiltonian.

ISBN 9780170394994

6 (The answers shown below are not the only possible answers.)

a A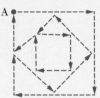

b A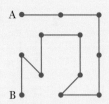

7 (The answers shown below are not the only possible answers.)

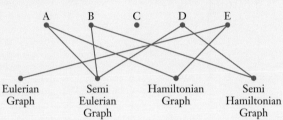
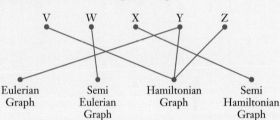

8

9

10 One possible route is: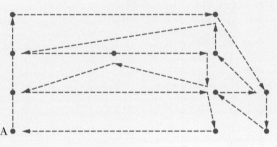

11 The graph is not an Eulerian graph so we cannot find an Eulerian circuit starting at A and finishing at A. However there are semi-Eulerian trails starting at A and finishing at B, where B is as shown in the diagram on the right. The route should follow one of these trails and then the road between B and A can be travelled for a second time to return to A.

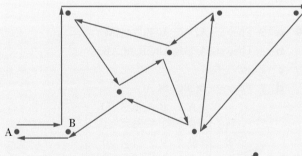

12 One of many possible circuits is shown on the right.

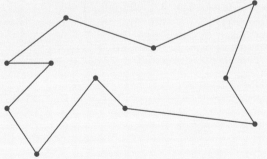

1

	A	B	C	D
A	0	1	1	1
B	1	0	1	0
C	1	1	0	0
D	1	0	0	0

2

	A	B	C	D
A	0	1	1	2
B	1	1	0	0
C	1	0	0	1
D	2	0	1	0

3

	A	B	C	D	E
A	0	0	0	1	1
B	0	0	0	1	1
C	0	0	0	1	1
D	1	1	1	0	0
E	1	1	1	0	0

4

	A	B	C	D
A	0	1	0	0
B	0	0	1	0
C	0	0	0	1
D	1	0	0	0

5

	A	B	C	D
A	0	1	0	1
B	0	0	0	1
C	0	0	0	0
D	2	0	0	0

6

	A	B	C
A	0	1	2
B	1	0	0
C	0	1	1

7

K_3
$$\begin{bmatrix} 0 & 1 & 1 \\ 1 & 0 & 1 \\ 1 & 1 & 0 \end{bmatrix}$$

K_4
$$\begin{bmatrix} 0 & 1 & 1 & 1 \\ 1 & 0 & 1 & 1 \\ 1 & 1 & 0 & 1 \\ 1 & 1 & 1 & 0 \end{bmatrix}$$

8

	A	B	C	D	E	F	G	H
A	0	1	1	1	0	0	0	0
B	1	0	1	0	0	0	0	0
C	1	1	0	1	1	0	1	0
D	1	0	1	0	0	0	0	0
E	0	0	1	0	0	0	0	0
F	0	0	0	0	0	0	1	0
G	0	0	1	0	0	1	0	1
H	0	0	0	0	0	0	1	0

9 a 3. They are ADAC, ACBC, ACAC.

b 3. They are ADAB, ABCB, ACAB.

c Zero.

d 2. They are BCBC, BCAC.

$M =$
	A	B	C	D
A	0	1	1	1
B	0	0	1	0
C	1	1	0	0
D	1	0	0	0

$M^3 =$
	A	B	C	D
A	1	**3**	**3**	2
B	**0**	1	**2**	1
C	3	2	1	0
D	2	1	1	0

10 a

$$
X = \begin{array}{c} \\ A \\ B \\ C \\ D \\ E \end{array}
\begin{array}{ccccc} A & B & C & D & E \end{array}
\left[\begin{array}{ccccc}
0 & 1 & 1 & 0 & 0 \\
1 & 0 & 1 & 0 & 0 \\
1 & 1 & 0 & 1 & 1 \\
0 & 0 & 1 & 0 & 0 \\
0 & 0 & 1 & 0 & 0
\end{array}\right]
$$

b

$$
X + X^2 = \begin{array}{c} \\ A \\ B \\ C \\ D \\ E \end{array}
\begin{array}{ccccc} A & B & C & D & E \end{array}
\left[\begin{array}{ccccc}
2 & 2 & 2 & 1 & 1 \\
2 & 2 & 2 & 1 & 1 \\
2 & 2 & 4 & 1 & 1 \\
1 & 1 & 1 & 1 & 1 \\
1 & 1 & 1 & 1 & 1
\end{array}\right]
$$

Matrix X shows the single flight connections between two of the cities.

Matrix X^2 shows the connections that can be done in two flights.

Thus if there are no zeros in matrix $X + X^2$ then every one of the five cities can be connected to each of the others by taking one or two flights. This being the case we can conclude that it is possible to fly with this airline between any two of these five cities in, at most, two flights.

11 a The fact that there is symmetry across the leading diagonal tells us that there are no 'one way only' connections.

  b Viewing the matrix $X + X^2 + X^3$, the presence of some zeros indicates that some hubs cannot be linked using 1, 2 or 3 direct links.

  Viewing the matrix $X + X^2 + X^3 + X^4$, the complete absence of zeros indicates that all hubs can be joined using 1, 2, 3 or 4 direct links.

  Thus all connections between two hubs can be done in up to four inter-hub connections, and some pairs do need four direct connections to connect them, for example A to J which could be linked using the path A-F-E-G-J.

  c Graph not shown here.

Miscellaneous exercise five PAGE 144

1 a Traversable. **b** Not traversable. **c** Traversable. **d** Traversable.
  e Traversable. **f** Not traversable. **g** Traversable. **h** Traversable.

Networks g and h are Eulerian.

Networks a, c, d and e are semi-Eulerian.

2 In a simple interest situation an initial amount, P, earns the same amount of interest, say I, each year.

  Thus, initial value   = P
  Value after 1 year   = $P + I$
  Value after 2 years   = $P + 2I$
  Value after 3 years   = $P + 3I$,   etc.

The constant increase from one term to the next is the characteristic of an arithmetic sequence.

The sequence of balances for simple interest is an example of an arithmetic sequence.

3 In a compound interest situation an initial amount, P, earns $r\%$ interest each year.

  Thus, initial value   = P
  Value after 1 year   = $P \times (1 + r/100)$
  Value after 2 years   = $P \times (1 + r/100)^2$
  Value after 3 years   = $P \times (1 + r/100)^3$ etc.

The constant multiplication from one term to the next is the characteristic of a geometric sequence.

The sequence of balances for compound interest is an example of a geometric sequence.

4 $T_1 = 5$, $T_2 = 17$, $T_3 = 53$, $T_4 = 161$, $T_5 = 485$.

5 $T_1 = 5$, $T_2 = 13$, $T_3 = 29$, $T_4 = 61$, $T_5 = 125$.

6

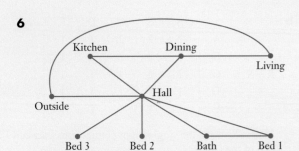

7 Without knowing more about the particular species of spiders, birds, snakes etc that frequent the area it is difficult to be sure exactly who might eat who. Hence the food web shown on the right is just a possible suggestion. There are spiders that eat birds, for example, and some spiders can eat snakes, so although the suggested answer on the right does not include these two particular examples of 'who is being eaten by who' it would not necessarily be wrong to include such directed edges.

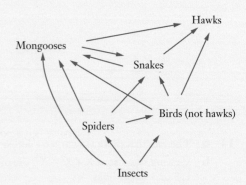

8 a $22\,000$ **b** $27\,000$

9 At the end of 32nd week the number of fish in the lake will fall to below 8000.

   Hence it will take 32 weeks (approximately, given that initial value of 10 000 is only approximate and the 1.5% might vary somewhat).

10 a $0.91, S = 1.23P - 34.9$ **b** 193 cm   **c** $P = 0.67S + 53.7$ **d** 178 cm

11 Allow others in your class to read and comment upon your report and you read and comment on theirs.

Exercise 6A   PAGE 152

1 105 units, HBW.   **2** 105 units, HBW.   **3** 95 units, HACW.

4 90 units, ABEFZ.   **5** 120 units, ACZ.   **6** 46 units, ACGZ.

7 125 units, ADFHJKMZ.   **8** 125 units, ABCDFHIZ.   **9** 60 minutes, ABCEGIK.

10 Ayesville – Buckeridge – Devizes – Ensmere – Hawton. $27 000 000.

11 a Meckering – Cunderdin – Quairading – Corrigin – Bullaring – Wickepin. 187 km.

   b Meckering – York – Beverley – Brookton – Pingelly – Narrogin – Wagin. 233 km.

12 The quickest journey from A to H takes 95 minutes and is along the route ACEFH.

13 Cataby – Badgingarra – Eneabba – Three Springs – Morawa. 214 km.

14 100 units, A – B – C – E – F – I.

15 Current Position – Waterfall – Derelict Cottage – Tourist Office – Fire Lookout – Next Campsite.

   4 h 50 mins (= 260 minutes + 30 minutes for lunch break).

   Current Position – Waterfall – Derelict Cottage – Tourist Office – Parker's Lookout – Tourist Office – Fire Lookout – Next Campsite.

   6 h 10 mins (= 340 minutes + 30 minutes for lunch break).

16 The chosen route should be MCFH for a total cost of $50 000.

   The requirement to go through junction S does change the minimum cost. The new route now needs to be MBSDH for a total cost of $53 000.

17 a AFGK   **b** ABEHIK   **c** ABEIK

ISBN 9780170394994

1

	A	B	C	D
A	0	0	0	1
B	0	0	0	1
C	0	0	0	1
D	1	1	1	0

2

	A	B	C
A	1	0	1
B	0	0	2
C	1	2	0

3

	A	B	C	D	E
A	0	1	0	0	1
B	1	0	1	0	0
C	0	1	0	1	0
D	0	0	1	0	1
E	1	0	0	1	0

4

	A	B	C	D
A	0	1	0	0
B	0	0	0	1
C	1	0	0	1
D	1	0	1	0

5

	A	B	C
A	0	1	0
B	0	0	1
C	1	0	0

6

	A	B	C	D	E
A	0	0	0	1	0
B	0	0	0	1	0
C	0	0	0	1	1
D	0	0	0	0	0
E	0	0	0	0	0

7 From the graph we can see that it is not possible to travel from one vertex to another (including back to itself) using exactly two edges. Hence the square of the adjacency matrix will be the 5×5 zero matrix.

8 a Head, tail, tail, head, head.

9 a The graph has exactly two odd vertices, vertex A (order 1) and vertex E (order 3). Thus whilst it is possible to find a trail that travels every edge once and once only, such a trail will start at one of the vertices A or E and finish at the other, for example the trail ABCDBEDE.

The existence of such a trail, i.e. one that starts at one vertex, finishes at a different vertex, and travels along each and every edge once only makes the graph semi-Eulerian.

b If we add edge AE to the graph all vertices will be even. We can then find a trail that will start and finish at the same point and pass along each and every edge once only. For example ABCDBEDEA. The existence of such a closed trail means that the graph will then be Eulerian.

10 a ABDE, 12 units.   ABDCE, 11 units.   ABCDE, 19 units.   ABCE, 12 units.

ACDE, 19 units.   ACE, 12 units.   ACBDE, 28 units.

b ABDCE.

11 CD, DE, GH, JK, KL, ML.

12 There are two equally shortest paths, each of length 23 units: ABCDFGH and ABCDEH.

13 a Approximately 81%. ($r^2 = 0.8098$)   **b** Approximately 24%. ($r^2 = 0.2419$)

14 $T_1 = 4$, $T_{n+1} = T_n + 1$.

15 a 7   **b** 17   **c** 2560   **d** 3584

16 Only **b** and **c**:   **b**

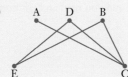

c

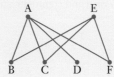

17 a 0.85   **b** $M = 0.95E - 11.07$ **c** 51

18 Cheapest upgrade will cost $144 000 and will be ACDEHJLZ.

19 a All three are connected graphs.

b Only the middle of the three (the tetrahedron) is a complete graph.

c Only the last one, the octahedron, is an Eulerian graph.

d All three graphs are Hamiltonian graphs.

20 100 minutes. Passing through Barton, Dane, Grange.

21

	A	B	C	D	E	Totals
Stack College	77	93	60	9	2	241
Bosun College	24	52	98	14	8	196
Totals	101	145	158	23	10	437

Examining row percentages:

	A	B	C	D	E	Totals
Stack College	32%	39%	25%	4%	1%	101%*
Bosun College	12%	27%	50%	7%	4%	100%

*Rounding anomaly.

As we move from one College's results to the other the proportion of students achieving each grade changes markedly. 32%, almost one third, of all the Stack College students achieved an A grade compared with 12%, about one eighth, of the Bosun College students. The proportions achieving B grades were also strongly in favour of Stack College, 39%, compared to Bosun College, 27%.

The changes in proportions achieving each grade as we move from the results of the students from one college to the other indicate that there is an association between the grades achieved and the college attended.

22 a There are six Hamiltonian circuits that can start at A and finish at A, though these really are three pairs of circuits with each pair consisting of the same circuit travelled in opposite directions.

The Hamiltonian circuits are:

ACBDA and ADBCA each with a total estimated time of 259 minutes.
ABCDA and ADCBA each with a total estimated time of 246 minutes.
ABDCA and ACDBA each with a total estimated time of 241 minutes.

Thus the Hamiltonian circuit given in the question, ACBDA, is not the quickest. The circuit ABDCA, travelled as stated or as ACDBA, gives the quickest time of 241 minutes.

b If we are only at A at start and finish and in between we visit each of B, C and D exactly once then it will be a Hamiltonian circuit that gives the shortest time. However, if multiple visits to the other vertices (and repeated trips along edges) are allowed it could be that the times make a non Hamiltonian path quicker. For the times shown on the right for example, and with repeat visits to any of B, C and D allowed, the quickest journey would be to avoid the three time consuming edges and follow ADBDCDA (or ADCDBDA) for a total of 30 minutes.

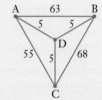

23 The withdrawal of $55 000 can last for ten years. At the end of the eleventh year the final withdrawal would be $16 916 (nearest dollar).

24 Allow others in your class to read and comment on your report and you read and comment on theirs.

INDEX

adjacency matrices 140–3
adjacent vertices 119, 122, 129
arcs 116, 122
arithmetic sequences (arithmetic progressions) 82–4, 95, 101
   growth and decay 95
   jumping to later terms 91–3
   recurrence rules 101–2
association between variables 6, 20

bipartite graphs 129
bivariate data viii, 2–27, 31–59
   trimming or cropping 52
bridge 121, 123, 129

categorical data viii
categorical variables viii–ix
cause and effect 41
closed path 118, 122
closed trail 118, 123
closed walk 118, 122
coefficient of determination 38, 54–5
coincidence 41
column graphs
   proportional 5, 6, 9, 13
   stacked 100% 5
column matrix xii
column percentages 5, 13–14
common difference (arithmetic sequence) 82
common ratio (geometric sequence) 85
complete graphs 121, 123
confounding variable 41
connected graphs/networks 119–121, 123, 136
connected vertices 119, 123
continuous data ix
continuous variables ix
coordinates x
correlation 20, 31–2

correlation coefficient 38–9
   calculation 42–3
   strong, moderate or weak relationship 42
covariance 43
cropping data 52
cycle 118, 122, 129, 136

data viii–ix
degree of a vertex 134
dependent variable 12, 21
diagonal matrix xii
difference rule for the sequence 68
digraphs 116–17, 122
dimensions of the matrix xii
directed edges 116, 122
directed graphs/networks 116–17, 120, 122, 134, 141
disconnected graphs/networks 119, 121, 123
discrete data ix
discrete variables ix

edges 108, 115, 122, 128
   directed 116
   multiple 115, 122
elements xii
equal matrices xiii
Eulerian circuit 136
Eulerian trail 136
Euler's rule 112, 128
even vertices 134
explanatory variable 12, 21
exponential graphs 85
exponential growth 95
extrapolation 22, 23, 33, 40

faces 108, 122, 128
first order linear recurrence relations 101–2

ISBN 9780170394994